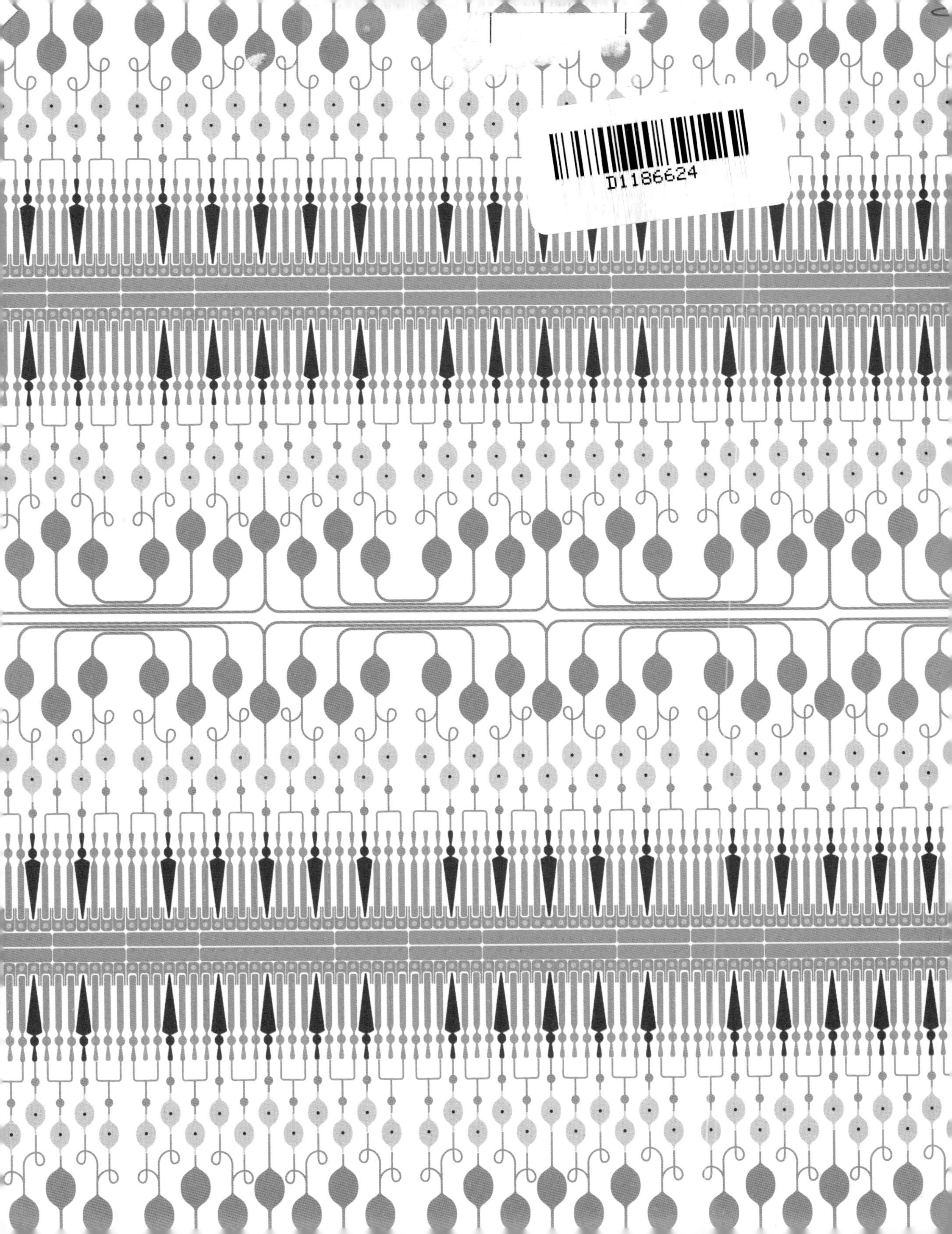
D1186624

BODY

A Graphic Guide to Us

Steve Parker is an author, editor and consultant specializing in information about the natural world, biology, technology and general sciences. Steve has a First Class Honours BSc in Zoology, is a Senior Scientific Fellow of the Zoological Society of London and has worked on the staff at London's Natural History Museum. He is the author of over 300 titles. For further information visit www.steveparker.co.uk.

Andrew Baker is an award winning illustrator with an international reputation, working in editorial, design and publishing contexts. Originally from Yorkshire, he studied at Liverpool and the Royal College of Art, and now lectures part-time at Middlesex University. He lives in East London with his wife Linda and their children Ray and James – whose contributions and support made these illustrations possible. Andrew is represented by Debut Art at www.debutart.com.

INTRODUCTION 4

PHYSICAL BODY

GENETIC BODY

CHEMICAL BODY

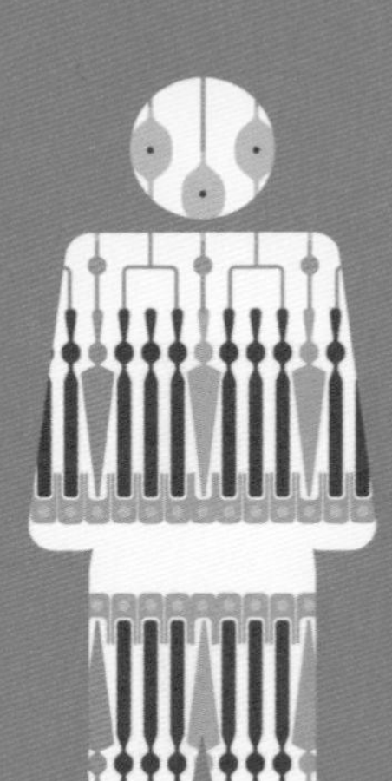

SENSITIVE BODY

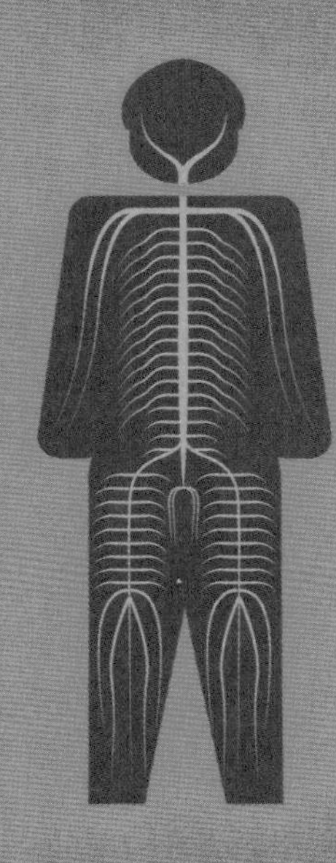

COORDINATED BODY

THINKING BODY

GROWING BODY

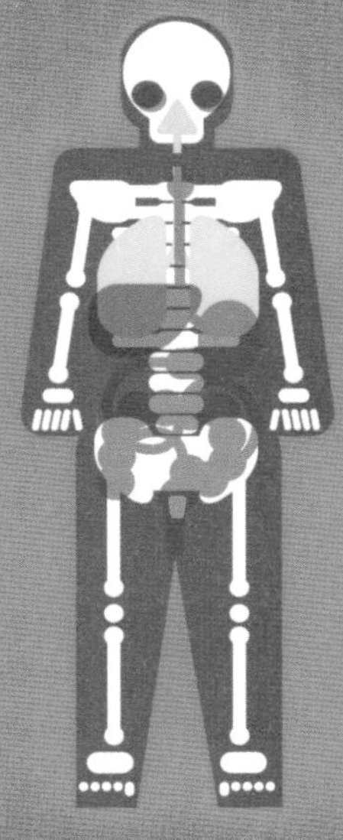

MEDICAL BODY

NOBODY IS JUST ANYBODY. EVERYBODY IS DEFINITELY SOMEBODY.

The most universal human possession works best when loved and cherished not only by its owner, but by closest family and dearest friends. Who wouldn't want to know absolutely everything about their human body, and even more besides?

Infographics are information and knowledge communicated in graphic form. Shapes and colours predominate over words and literacy. Infographics are instinctive to comprehend and fast to assimilate. They transcend language and easily spring back into the memory. They can be understood by anyone, someone and indeed everyone. They even make statistics fun, data entertaining, and knowledge stick like glue.

So it seemed an excellent idea to combine these two topics, human body and infographics. But how to organize it all? Many body books run through the dozen or so functional systems: bones, muscles, heart and blood, digestion, brain and nerves, and the rest. However, we wanted this book to be different.

Going back to Renaissance times and the birth of modern knowledge, the body was studied in two chief ways. One was anatomy: physical structure, materials and construction; kickstarted by the momentous *On the Fabric of the Human Body* by Andreas Vesalius in 1543. Anatomy's complementary partner is physiology: chemical workings and functions, introduced as a concept in 1567's *Physiologia* by Jean Fernel. These paired topics still form the basis of modern human biology and medicine – and sections 1 and 2 of this book. A belated latecomer is section 3, the genetic body. This has existed only since the mid-20th Century, marked especially by one of the greatest discoveries in all of science – the structure of DNA in 1953 by James Watson and Francis Crick.

The human body learns and experiences through its senses, and all the main sensory modalities are explored in section 4. Its parts – cells, tissues and organs – are also intensely coordinated and integrated into a unified whole, as explained in section 5. Towering over the whole living organism is its chief command-and-control centre, intranet hub, and seat of awareness, perceptions and consciousness – the brain, described in section 6. So far, all of this is in the adult. Every body has a history. It starts as a pinpoint-sized fertilized egg which increases billions of times in size and complexity, and section 7 tracks this life cycle aspect. And when things go wrong, medicine is ready to lend a hand, as described in section 8.

No body book can hope to be comprehensive. But being selective, fascinating, intriguing, surprising, individual, regional and global can work as stand-ins, especially using this graphic approach. Flowcharts, diagrams, maps, step-by-steps, timelines, symbols, isotypes, icons, pies and bars are all involved. For the basic material they feed on, we are indebted to those who measure, collate and analyze such vast amounts of raw data, bare facts and naked information. Our task has been to find, interpret and transform so that readers find something appealing to take away. Hopefully you will be encouraged to understand and appreciate your most treasured possession just that bit more.

PHYSICAL BODY

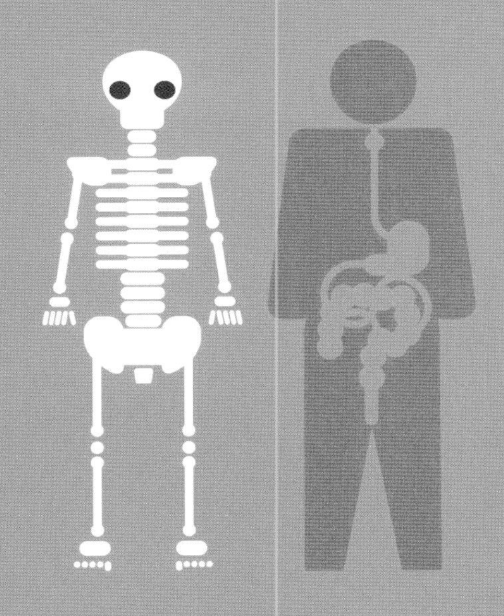

MILE-HIGH BODY

A typical human body is a massively complex ongoing interaction of dozens of organs, consisting of hundreds of tissue subtypes, themselves composed of billions of microscopic cells. One way to visualize this colossal complexity and enormous range of physical sizes is to enlarge the body to – let's say, a convenient 1 mile or 1.6 km in height. This is twice as tall as the world's loftiest skyscraper, at which scale humans themselves appear as mere ants swarming in, out, up and down – although see below!

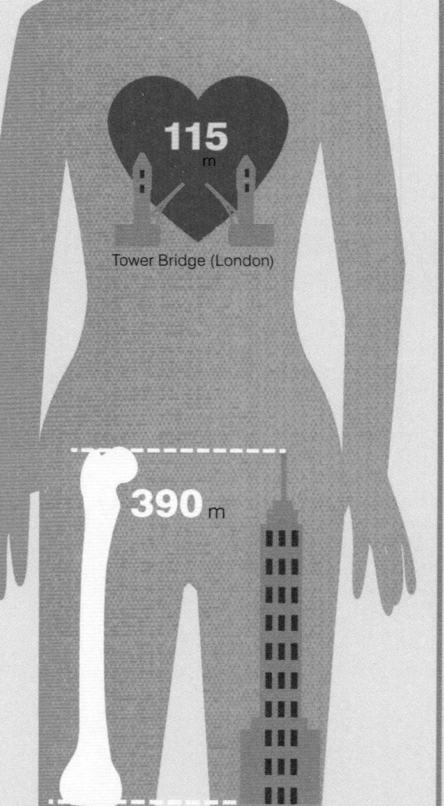

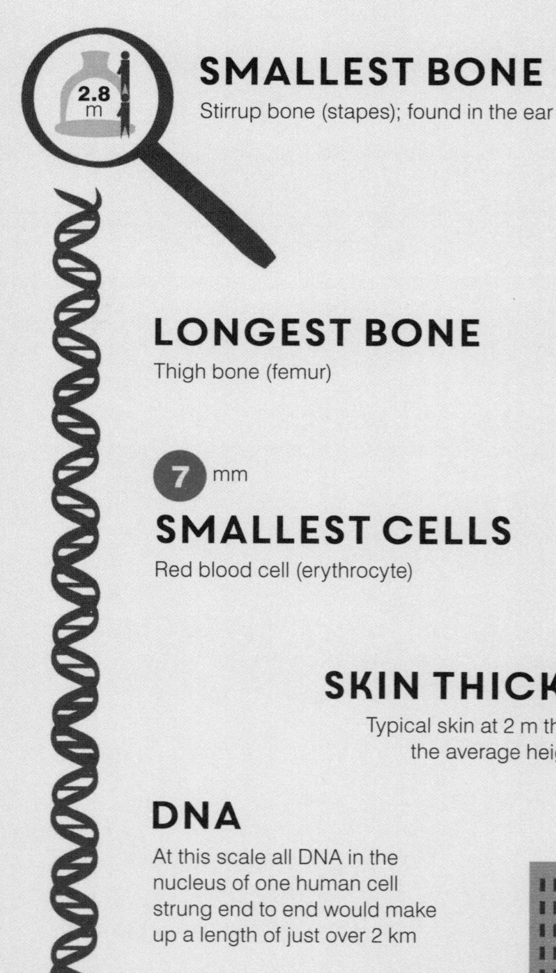

SMALLEST BONE

Stirrup bone (stapes); found in the ear

LONGEST BONE

Thigh bone (femur)

SMALLEST CELLS

Red blood cell (erythrocyte)

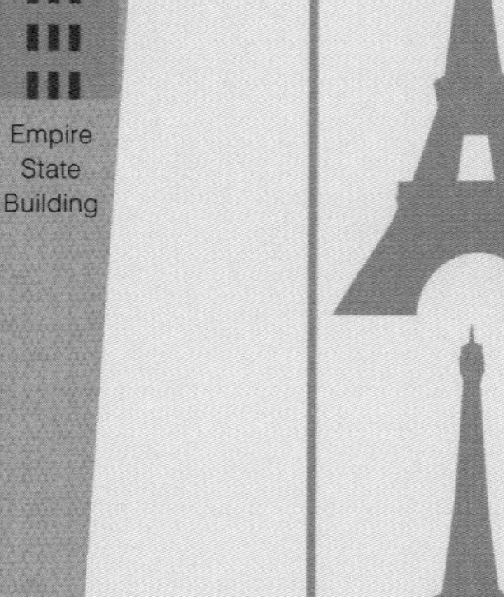

SKIN THICKNESS

Typical skin at 2 m thick, which is the average height of a door

DNA

At this scale all DNA in the nucleus of one human cell strung end to end would make up a length of just over 2 km

1 mile = 1.6 km / 1,600 m

The Eiffel Tower (Paris)

Ovum (egg cell) 11 cm

2 cm

White blood cell (macrophage)

EYELASH Length of a medium human hand

5 mm

TYPICAL CELL NUCLEUS

DNA in the nucleus
2 μm (micrometres) diameter

1/30 width of a human hair

1/60 thickness of this page

WALKING TALL

Height or stature is probably the most easily visualized measure of the human body. Average heights have been increasing worldwide for at least two centuries, mainly due to better nutrition – especially during infancy – coupled with less illness and disease. This trend is most progressive in advanced or richer countries. Currently it is very marked in the Netherlands, where young adult males average 184 cm and females 170 cm in height – some 19 cm taller than their counterparts 150 years ago. Yet in North America, average heights have increased only slightly since the mid 20th Century. Worldwide, heights are likely to increase for several decades yet. If nutrition and general health improve for poorer countries, their averages will rise relatively fast, while those in the richer regions seem to be gradually reaching a plateau.

Preceding human species

600,000–250,000 years ago
Homo heidelbergensis
(Europe, Africa)

200,000–50,000 years ago
Homo neanderthalensis
(Europe, Asia)

3,200 years ago
(Ancient Greece)

Mid 10th Century
(Europe)

Mid 17th Century
(Europe)

Mid 18th Century
(Europe)

Mid 19th Century
(Europe, North America)

Mid 20th Century
(Western Hemisphere)

SOME NOTABLE AVERAGE HEIGHTS

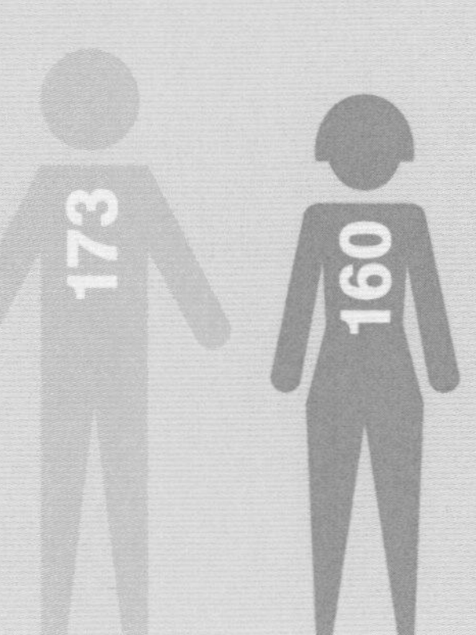

World

Batawa pygmy people (Africa)

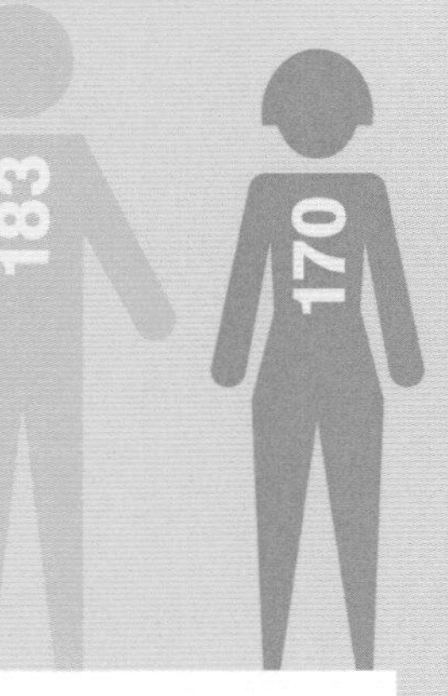

Dinka people (Africa)

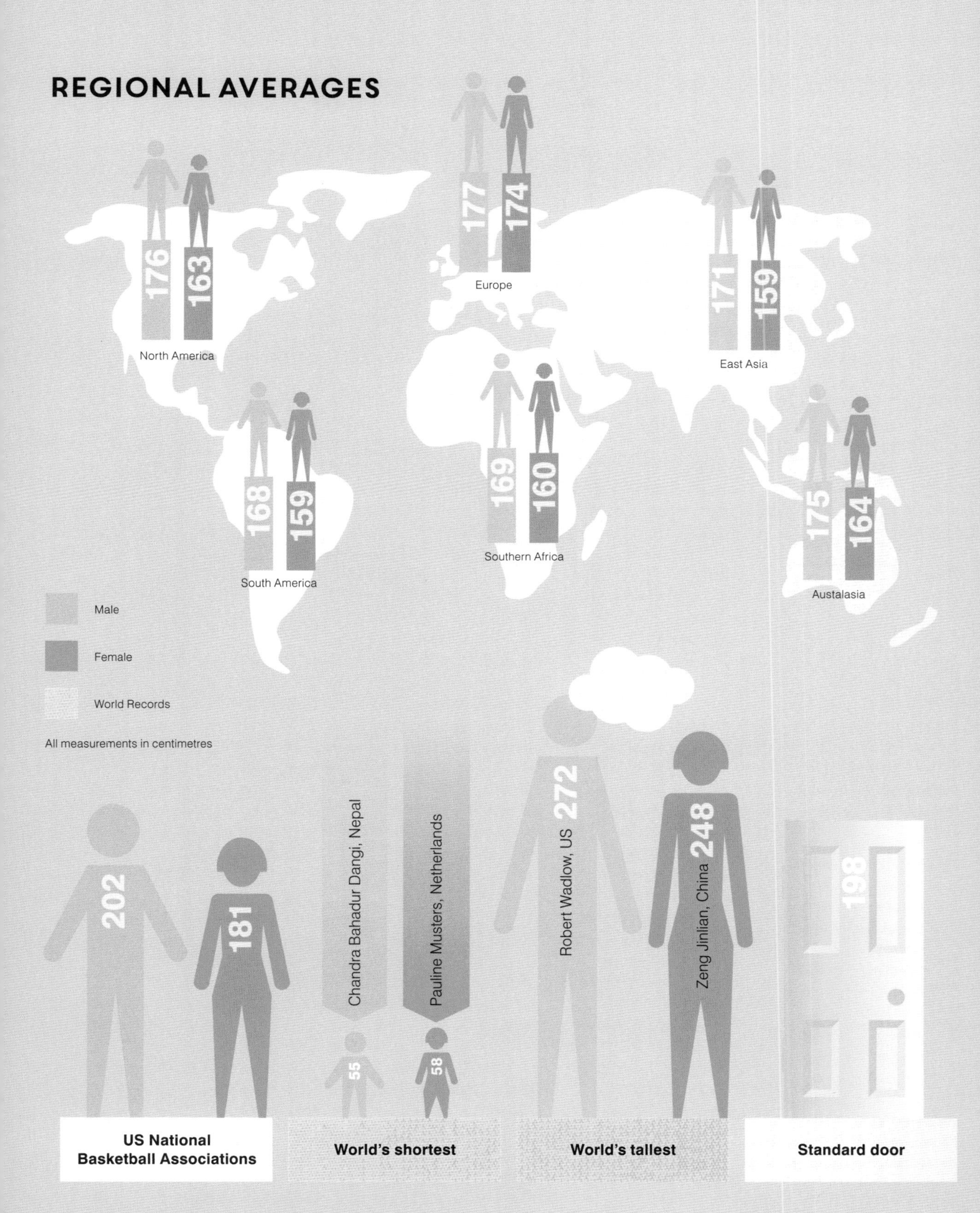
REGIONAL AVERAGES
176
163
North America
177
174
Europe
171
159
East Asia
168
159
South America
169
160
Southern Africa
175
164
Austalasia
Male
Female
World Records
All measurements in centimetres
202
181
US National Basketball Associations
Chandra Bahadur Dangi, Nepal
55
Pauline Musters, Netherlands
58
World's shortest
Robert Wadlow, US
272
Zeng Jinlian, China
248
World's tallest
198
Standard door

BODY BUILDS

All human skeletons have 206 bones (apart from rare instances of unusual development or surgical removal). But the relative sizes and shapes of bones vary from one owner to another, giving a variety of basic body builds: big-boned, svelte, long-limbed, chunky, solid, rangy, delicate and many other metaphorical descriptors.

After growth finishes, the adult skeletal shape determines body measures such as overall height and limb proportions. But the layers that clothe the skeleton also add greatly to the body's general outline. These include several sets of muscles, from deep to superficial, and on top of them, the overcoat of skin with its endlessly debated underlayer of hypodermal subcutaneous adipose tissue – fat.

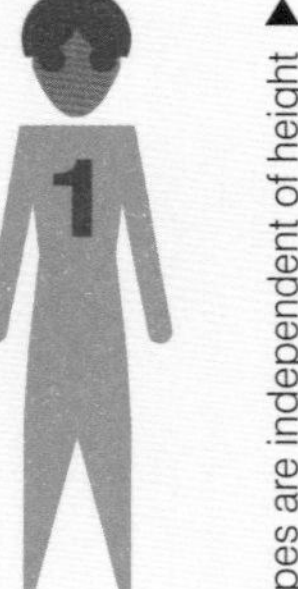

GENERAL SKELETAL TYPES

1 Ectomorph: Slim, light-boned, 'gracile', tendency to leanness

2 Mesomorph: Average

3 Endomorph: Wide, heavy-boned, 'robust', tendency to stoutness

Most individuals are combinations of two types.

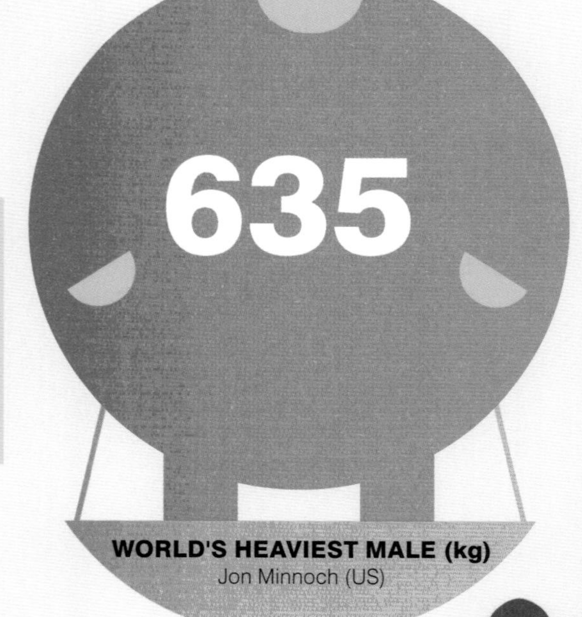

In the 1940s US psychologist William Sheldon attempted to correlate body shape and size to personality traits, temperament, intelligence and emotional states, e.g. ectomorphs being introverted, anxious, shy and restrained; endomorphs being open, expressive, voluble and easy-going. This theory has since been discredited.

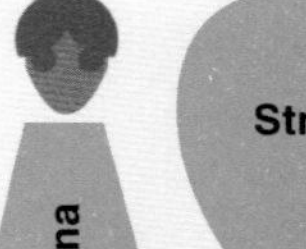

healthy body proportions (% mass)

Muscle

Bone

Other

Fat

Male: 15, 45, 15, 25

Female: 12, 35, 28, 25

FRUIT'N'NUT BODIES

Fruity or nutty body shapes may be easier to remember than complicated formulae. These portray where extra weight is carried. In general, having abdominal fat (an apple shape) indicates higher health risks than carrying fat around the buttocks and thighs (a pear shape).

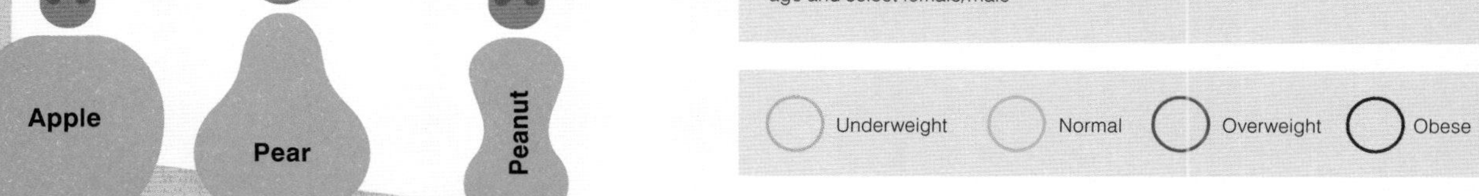

BMI: body mass index

BMI is an approximation designed to relate mass (weight), height and health implications. It aims to encompass both females and males, and to account for most body shapes, from slim to wide.

$M \div H^2$ or body's mass (weight) in kilograms divided by body height times itself in metres

WHtR: waist to height ratio

Probably the simplest mathematically for these types of calculations, WHtR is a quick-and-ready indicator of where body fat is distributed.

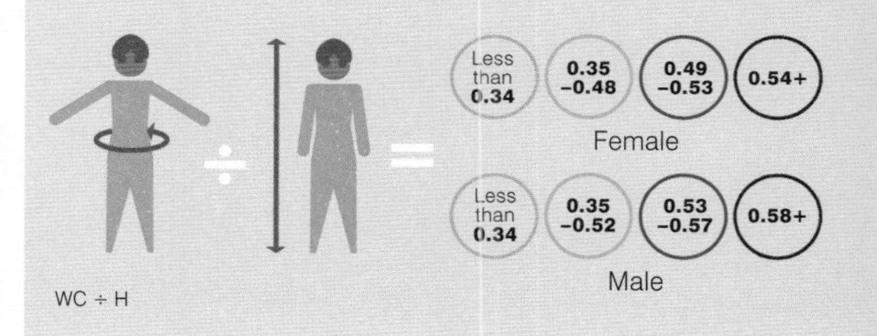

$WC \div H$

ABSI: a body shape index

A development of BMI, ABSI includes waist circumference (WC). It takes into account body fat distribution and is said to be a more accurate health predicting factor, however it involves trickier mathematics.

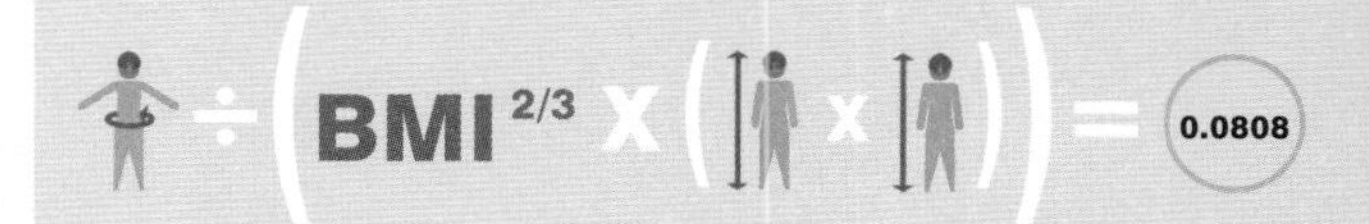

$WC \div (BMI^{2/3} \times H^2)$ or BMI to the power $^2/_3$ multiplied by body height times itself in metres, divided into waist circumference in metres. The full calculation also includes age and select female/male

Underweight

Normal

Overweight

Obese

IN PROPORTION

Since antiquity, artists and sculptors have celebrated proportion and harmony in the human form. Of course, bodies come in many shapes and sizes, but much the same proportions and relationships apply to the great majority of them. The well-known golden ratio of 1:1.618 (also variously known as the golden section, golden mean, phi, Φ) is seen extensively in nature and is much used in art to produce lengths and shapes with pleasing and balanced relationships. This ratio also occurs throughout the body.

1 1.618

Head Units ('Face Eighths')
Taking the head/face, from base of chin to crown of scalp, as one-eighth of total body height leads to these typical proportions:

The Golden Body
Golden ratio. For two lines lengths a and b a:b = (a+b):a = 1.618)

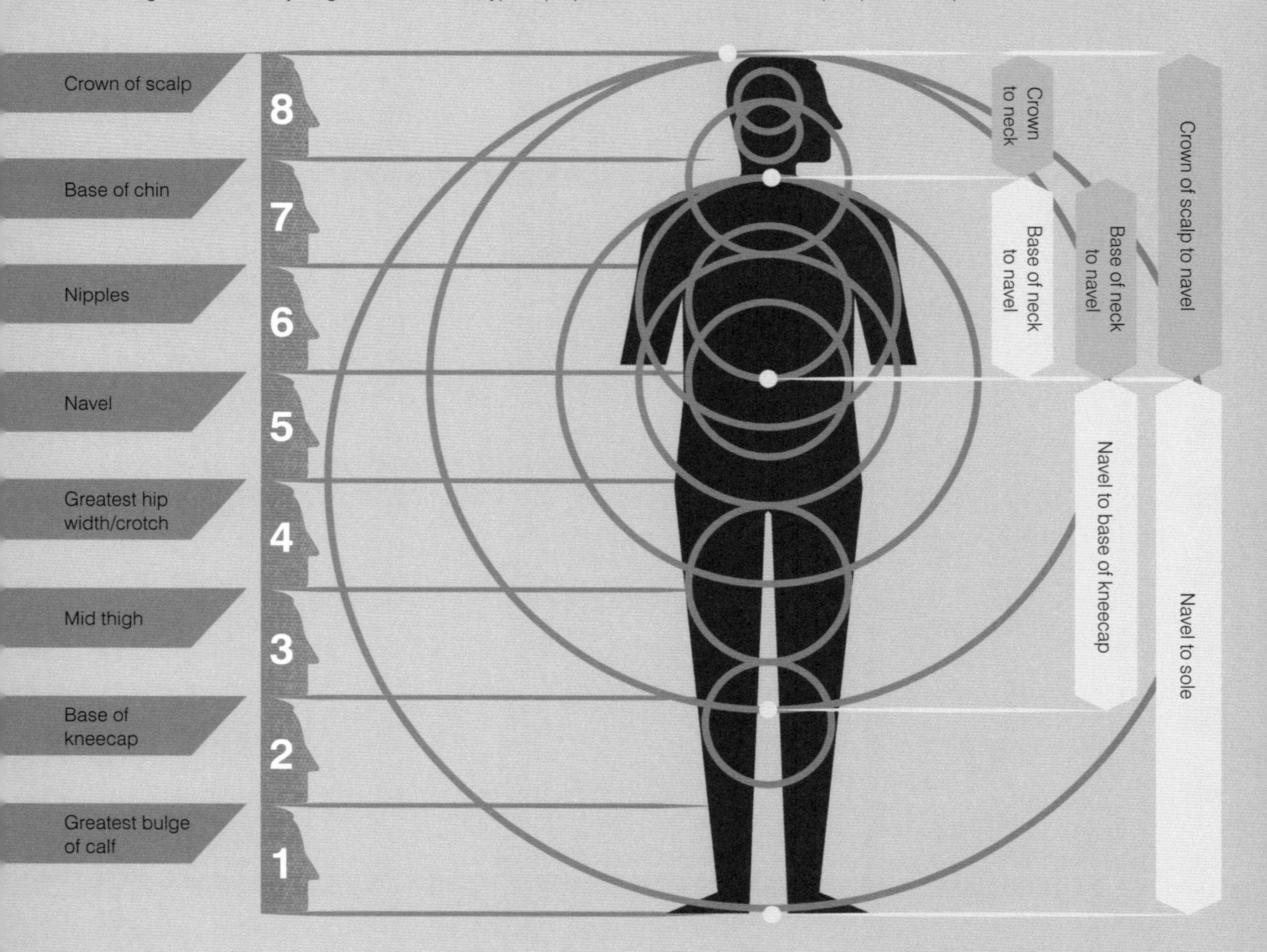

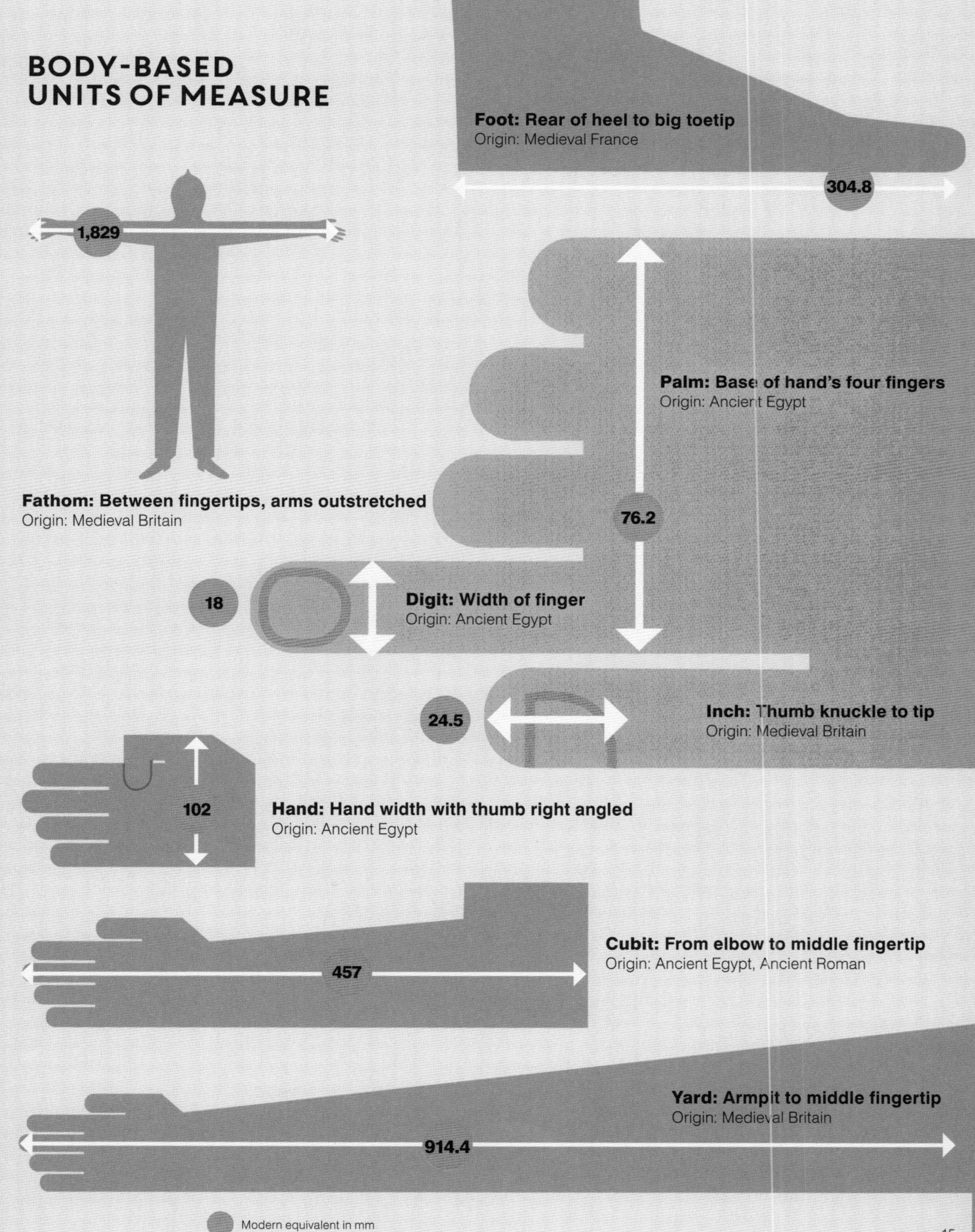
BODY-BASED UNITS OF MEASURE
Foot: Rear of heel to big toetip
Origin: Medieval France
304.8
1,829
Palm: Base of hand's four fingers
Origin: Ancient Egypt
Fathom: Between fingertips, arms outstretched
Origin: Medieval Britain
76.2
18
Digit: Width of finger
Origin: Ancient Egypt
24.5
Inch: Thumb knuckle to tip
Origin: Medieval Britain
102
Hand: Hand width with thumb right angled
Origin: Ancient Egypt
Cubit: From elbow to middle fingertip
Origin: Ancient Egypt, Ancient Roman
457
Yard: Armpit to middle fingertip
Origin: Medieval Britain
914.4
Modern equivalent in mm

SLICED AND DICED

Much as finding a place in the landscape requires latitude, longitude and altitude, precise location of any part of the body demands a basic set of three-dimensional coordinates, or matrices: up-down, side-side, front-rear. Today's plethora of scanning technologies reveal the body's insides like never before, without a scalpel in sight. These are the need to know directions when viewing the body.

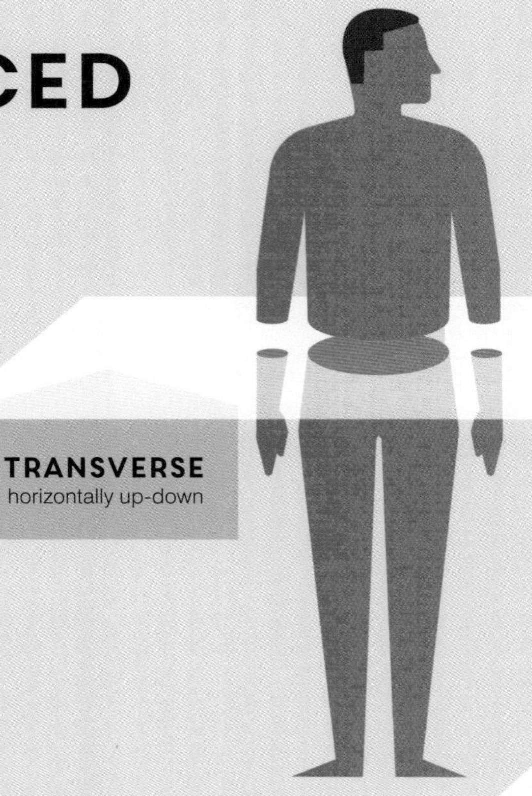

ANATOMICAL PLANES

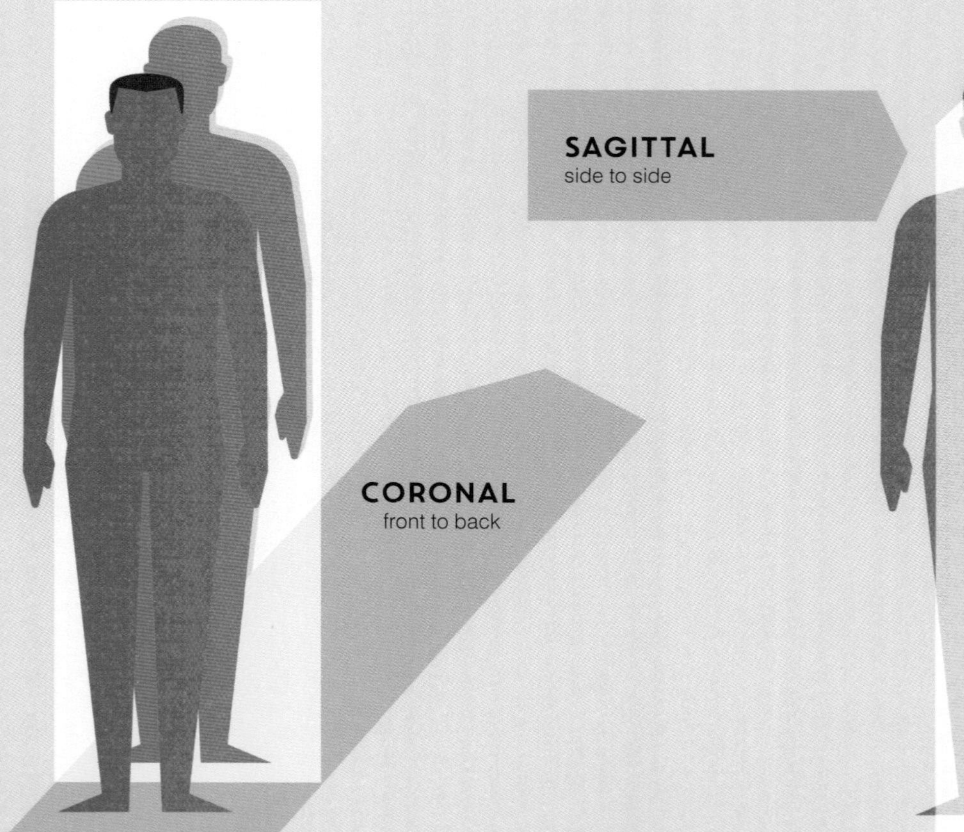

AXES OF ROTATION

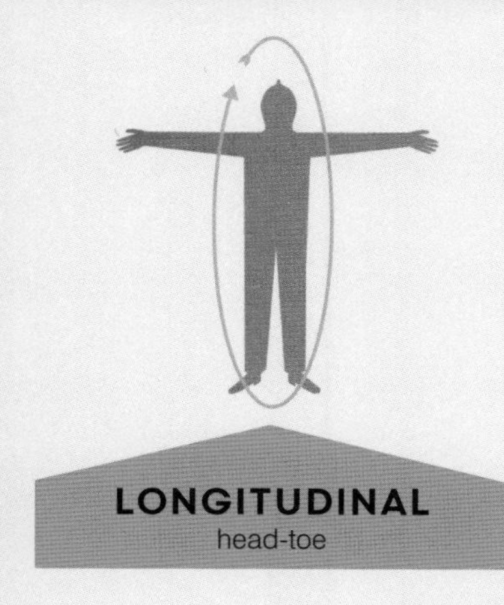

LONGITUDINAL
head-toe

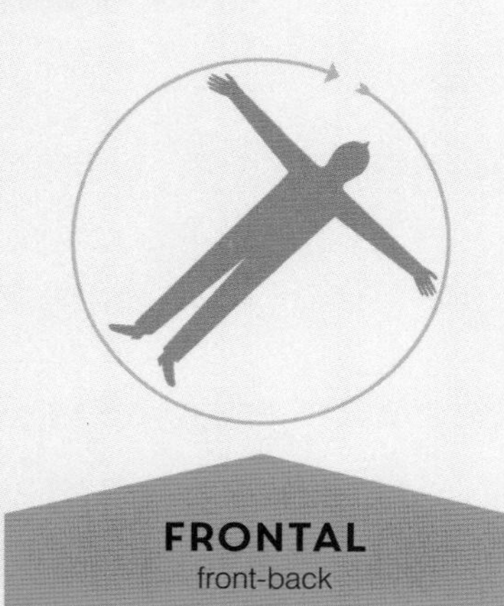

FRONTAL
front-back

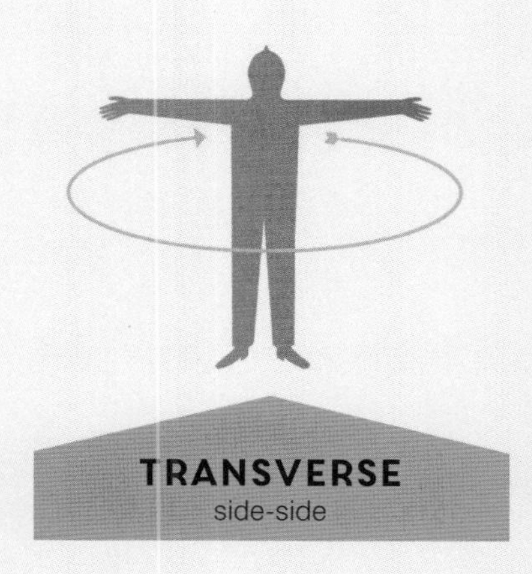

TRANSVERSE
side-side

DIRECTIONS OF VIEW

INFERIOR
from below

SUPERIOR
from above

LATERAL
sideways away from midline

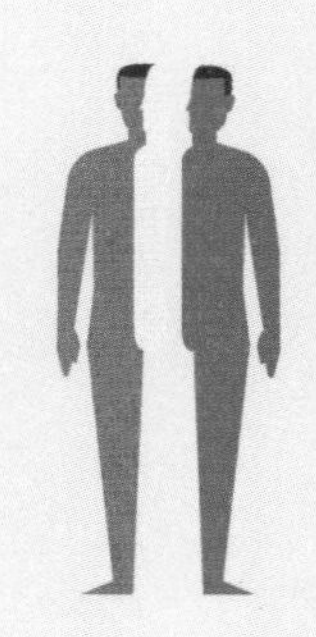

MEDIAL
sideways towards midline

ANTERIOR
from the front

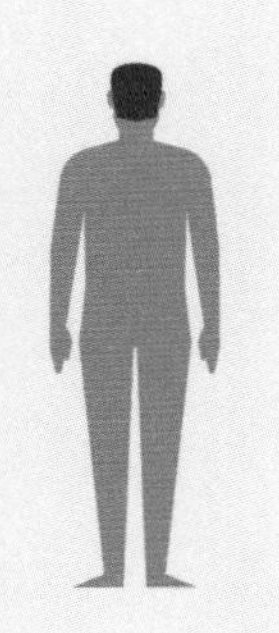

POSTERIOR
from the rear

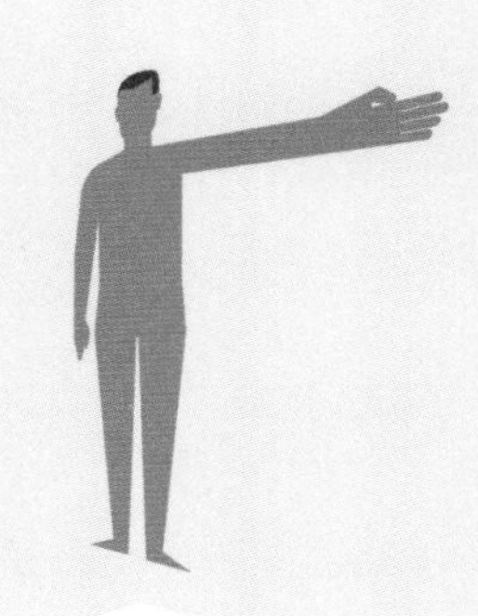

DISTAL
extremity towards main body

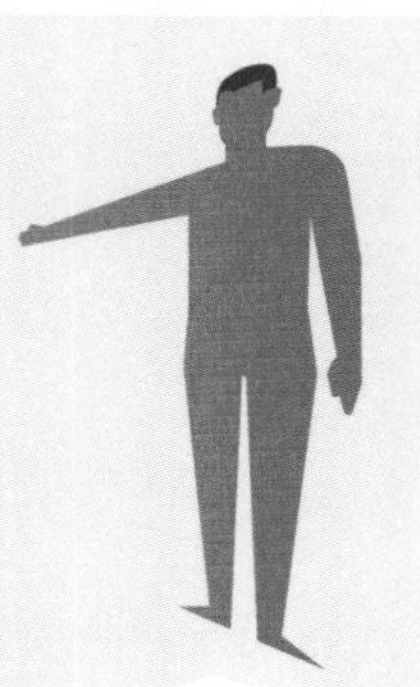

PROXIMAL
main body towards extremity

SEE-THROUGH BODY

Advanced imaging technologies mean we can see into and even right through the body, and visualise every nook and cranny, without even touching a scalpel. Here are some of the major organs and landmarks that provide good reference points, and which are all explained later in the book.

1 frontal bone
2 orbicularis oculi muscle
3 eye orbit
4 zygomatic muscle
5 nasal cavity
6 carotid artery
7 jugular vein
8 cervical lymph nodes
9 scapula
10 thyroid
11 axillary artery and vein
12 thoracic lymph nodes
13 thymus
14 humerus
15 heart
16 ribs
17 left lung
18 left kidney
19 aorta
20 gall bladder
21 liver
22 radial vein

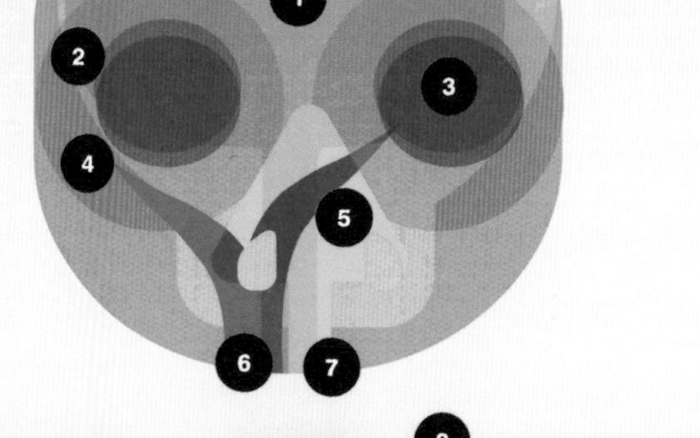

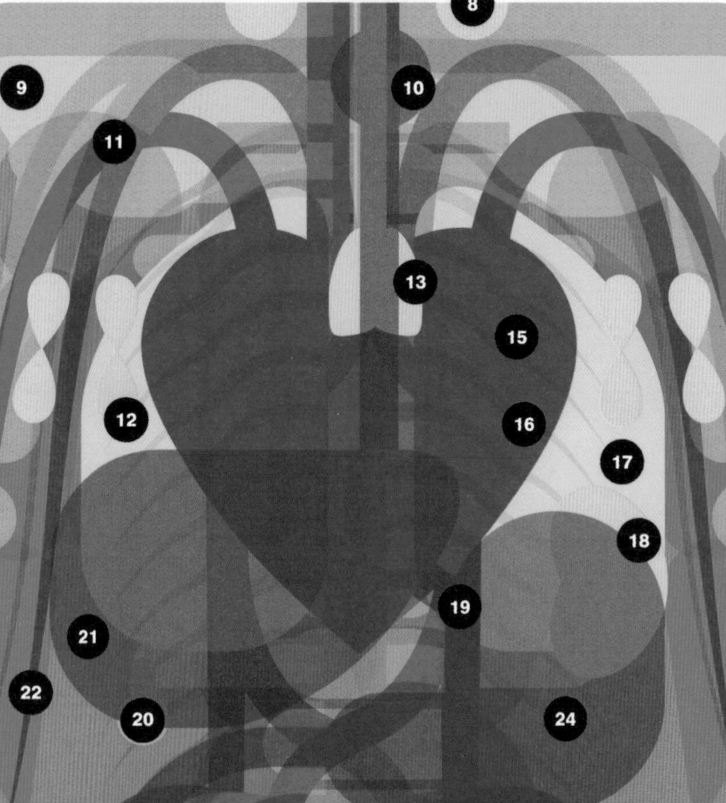

23 ulnar artery
24 stomach
25 radius and ulna
26 carpals
27 metacarpals
28 small intestine
29 colon (large intestine)
30 iliac artery
31 iliac vein
32 appendix
33 inguinal lymph nodes
34 rectum
35 femur
36 patella
37 femoral artery and vein
38 tibia and fibula
39 anterior tibial artery and vein
40 posterior tibial artery and vein
41 tarsals
42 metatarsals

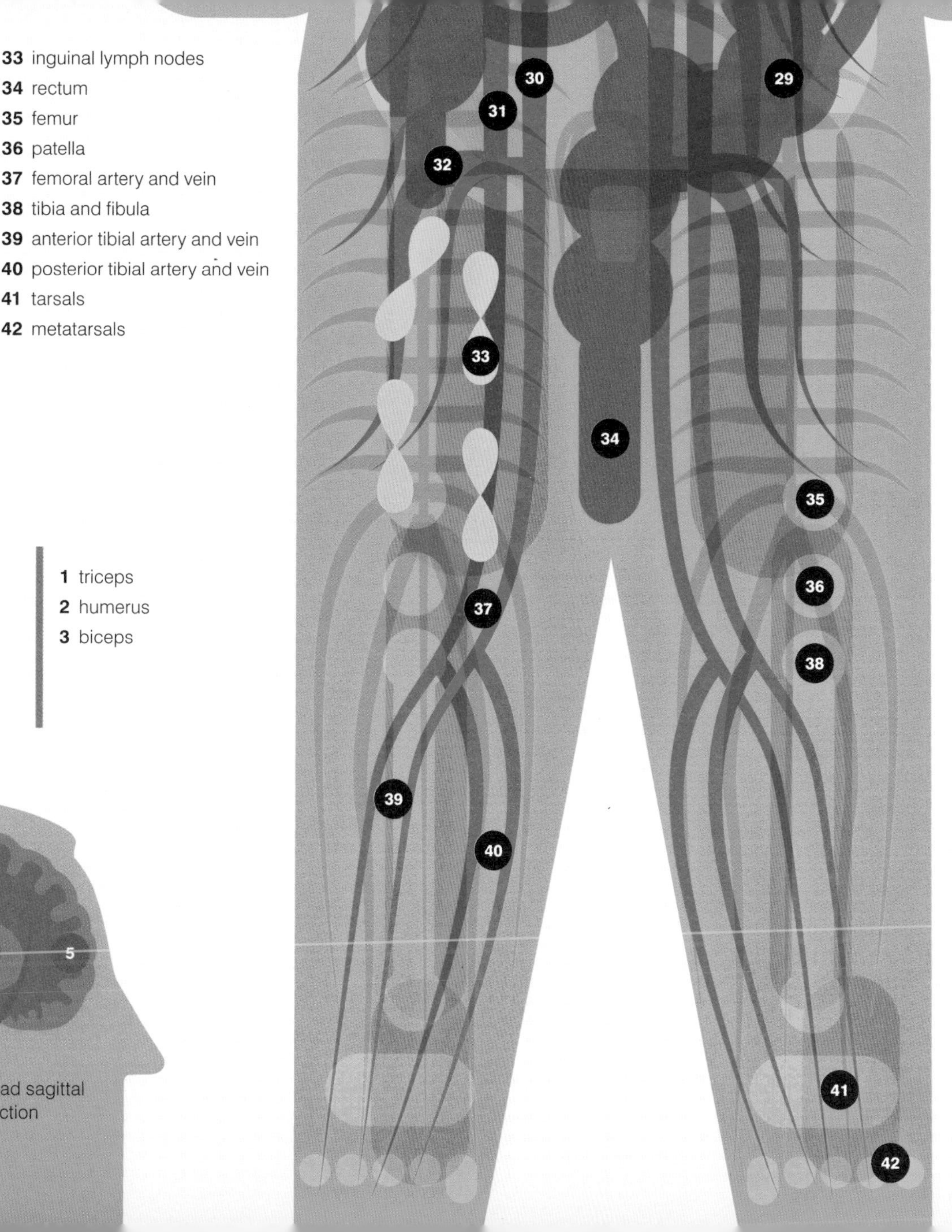

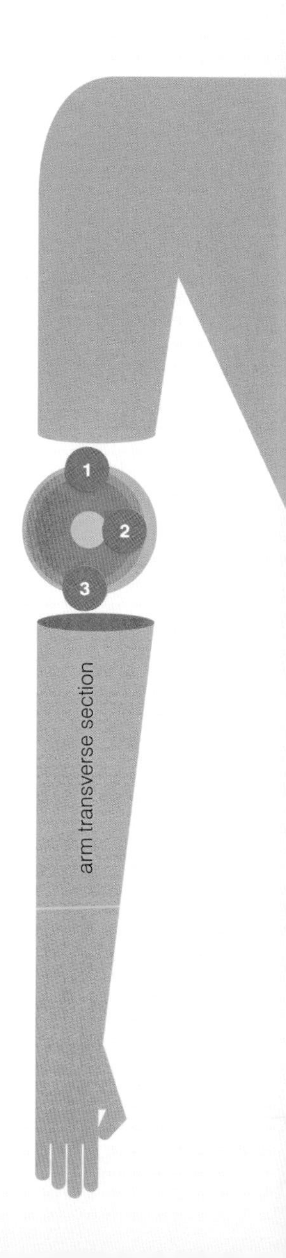

1 occipital lobe
2 cerebral cortex
3 cerebral ventricle
4 corpus callosum
5 fontal lobe

1 triceps
2 humerus
3 biceps

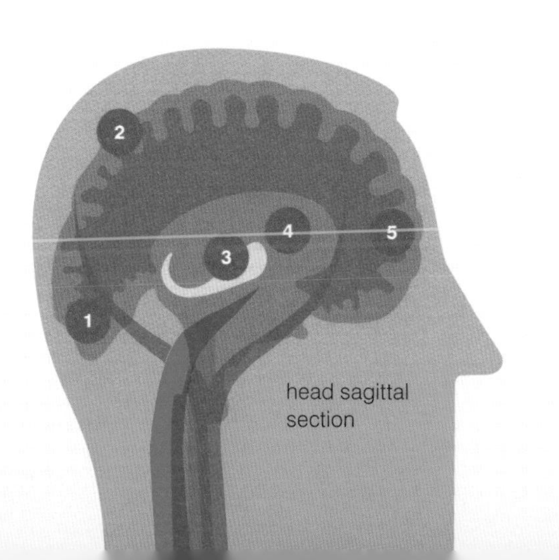

SYSTEMS ANALYSIS

A body system is a group of organs, tissues and cells all dedicated to one (perhaps two) major functions that keep human beings alive and working well.

Integumentary
• Skin • hair • nails • sweat and other exocrine glands
For protection, temperature control, waste removal and sensation.

Muscular
640 skeletal muscles specialized for contraction.
For bodily movement, movement of internal substances and protection.

Cardiovascular
• Heart • blood
• blood vessels
For delivery of oxygen and nutrients, collection of carbon dioxide and wastes, temperature control.

Urinary
• Kidneys • ureters
• bladder • urethra
For filtering wastes from blood and controlling overall levels of bodily fluids.

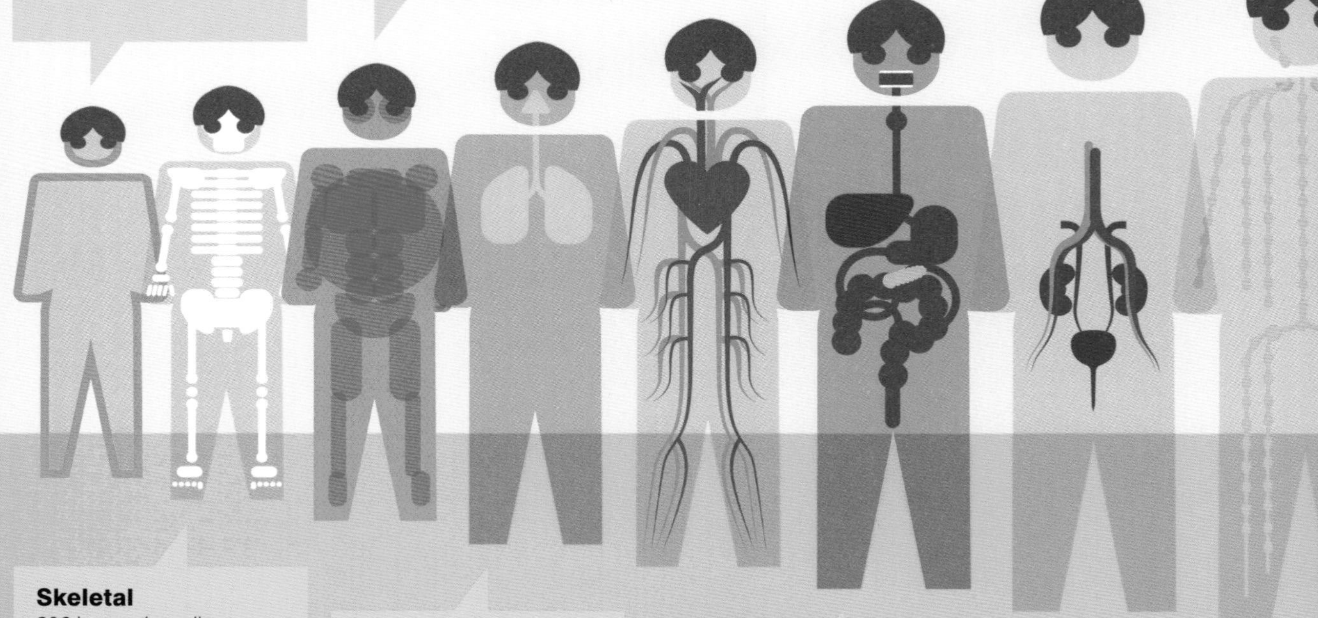

Skeletal
206 bones (usually also including joints)
For support, protection, movement and production of blood cells.

Respiratory
• Nose • throat • windpipe
• airways • lungs
For absorbtion of oxygen, removal of carbon dioxide, vocalization.

Digestive
• Mouth • teeth • salivary glands • oesophagus
• stomach • intestines
• liver • pancreas
For physical and chemical digestion and absorption of nutrients.

Lymphatic
• Lymph nodes • lymphatic vessels • white cells
For draining general body fluids, distributing nutrients, collecting waste and repairing and defending the body.

Immune

- White blood cells
- spleen • lymph nodes
- other glands

For body defences against germs and other invaders, cancers and other diseases.

Sensory

- Eyes • ears • nose • tongue
- skin • inner sensory parts

For information about surroundings (sight, hearing, smell), body position and movement, and internal conditions such as muscle tension, joint position, temperature, etc.

Reproductive

Female: ovaries • fallopian tubes • uterus • vagina • associated ducts and glands

Male: testes • penis • associated ducts and glands

For the production of offspring; this is the only system to differ between females and males and the only system not necessary for survival.

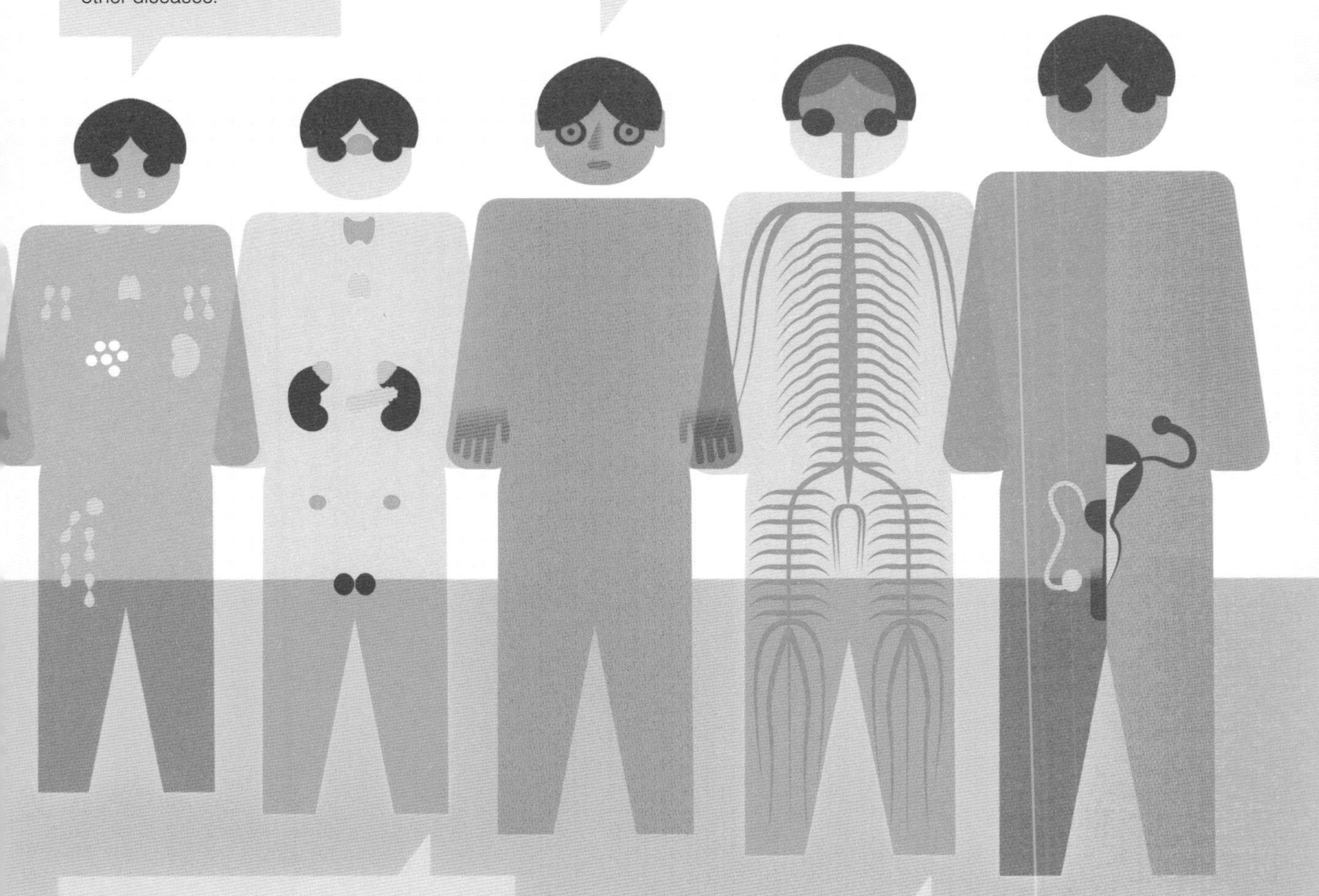

Endocrine

Hormonal (endocrine) glands such as pituitary, thyroid, thymus, adrenals

For production of chemical hormone substances for the communication and coordination for growth, digestion, fluid levels, fear reactions and many other processes.

Nervous

- Brain • spinal cord • nerves

For collecting and processing information, thoughts, decisions, memories and emotions, and for controlling muscles and glands.

PARTS MAKE A WHOLE

There are many ways of dividing up a human body. In terms of working roles or functions there are systems, organs, tissues, cells and their biochemical processes, or physiology. From the anatomical or structural perspective there are organs and tissues again, the largest being the skin (with its fatty or subcutaneous layer) and liver. Another anatomy-based approach is regional – the head, the torso comprising the chest or thorax above and the abdomen below, and the limbs with their various segments.

	BODY MASS %	MASS IN GRAMS IN A 75 KG BODY
Muscles	40	30,000
Skin (all layers)	15	11,200
Bones	14	10,500
Liver	2	1,550
Brain	2	1,400
Large intestine	1.5	1,100
Small intestine	1.2	900
Right lung	0.6	450
Left lung	0.5	400
Heart	0.5	350
Spleen	0.18	140
Left kidney	0.18	140
Right kidney	0.17	130
Pancreas	0.13	100
Bladder	0.1	75
Thyroid	0.05	35
Uterus (female)	0.08	60
Prostate (male)	0.03	20
Testis (male)	0.03	20

SIZING THE BODY

Traditional measures for body garments in the UK (inches).

Hat
Circumference around widest part (just above eyebrows) divided by 3.15.

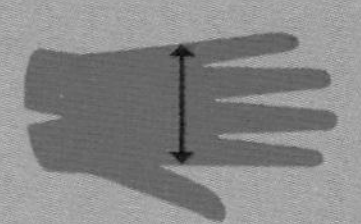

Glove
Around the widest part (knuckles).

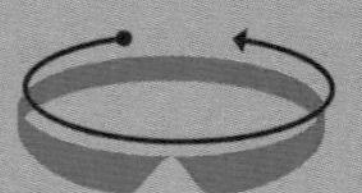

Collar
Around the fullest part of the neck plus ½ inch.

Sleeve
from centre back of neck to shoulder, then shoulder to wrist bone.

BIGFOOT TREND

Developed regions such as North America and Europe have seen a recent trend to larger feet, especially in women (as shown here by these average adult female sizes). Part of this is accounted for by increased stature – but not all.

1960 UK 4 Europe 37 US 6½

1970 UK 5 Europe 38 US 7½

2010 UK 6.5 Europe 39½ US 8½

Try for Comparison!

Shoe

Based on King Edward II's (1284–1327) foot being size 12 (12 inches), then decreasing or increasing by one barleycorn (⅓ inch) per size.

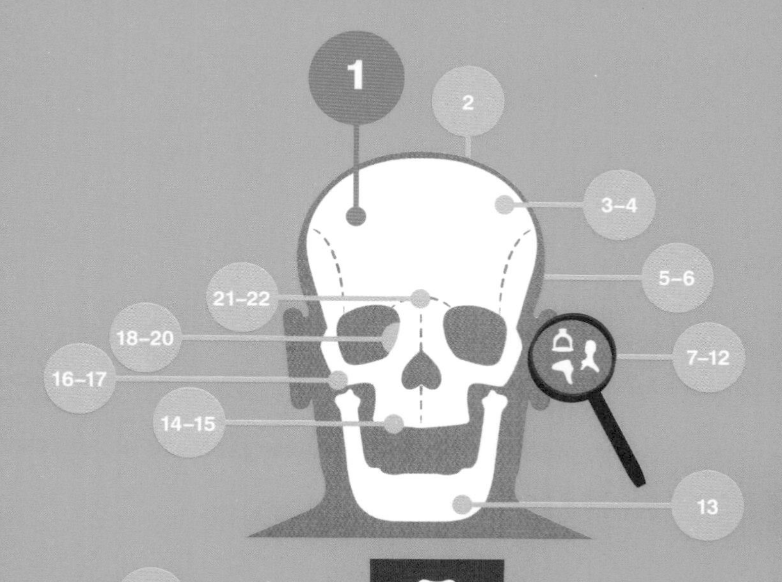

The human body has

206

bones (usually)

BARE BONES

During early development in the womb, bones form first as shapes of cartilage; gradually these shapes become ossified, or infiltrated with bone material, and the number of actual bones reaches more than 300 during infancy. Then the total falls again as some of these bony elements, especially those in the skull, join or fuse towards maturation.

Genetic and developmental variations occur too. About one person in 120 has two extra ribs, making 13 pairs rather than 12. About one in 25 skeletons are 'lumbarized', with what appears to be a sixth lumbar vertebrae added to the regular five; however, the extra one is a mobile non-fused vertebra 'borrowed' from the sacrum below, which has four fused vertebral segments rather than five. Also around one person in 100 has a variant in the numbers of fingers or toes and their bones. Then, sometimes there are the occasional extra bones in the wrist, ankle...

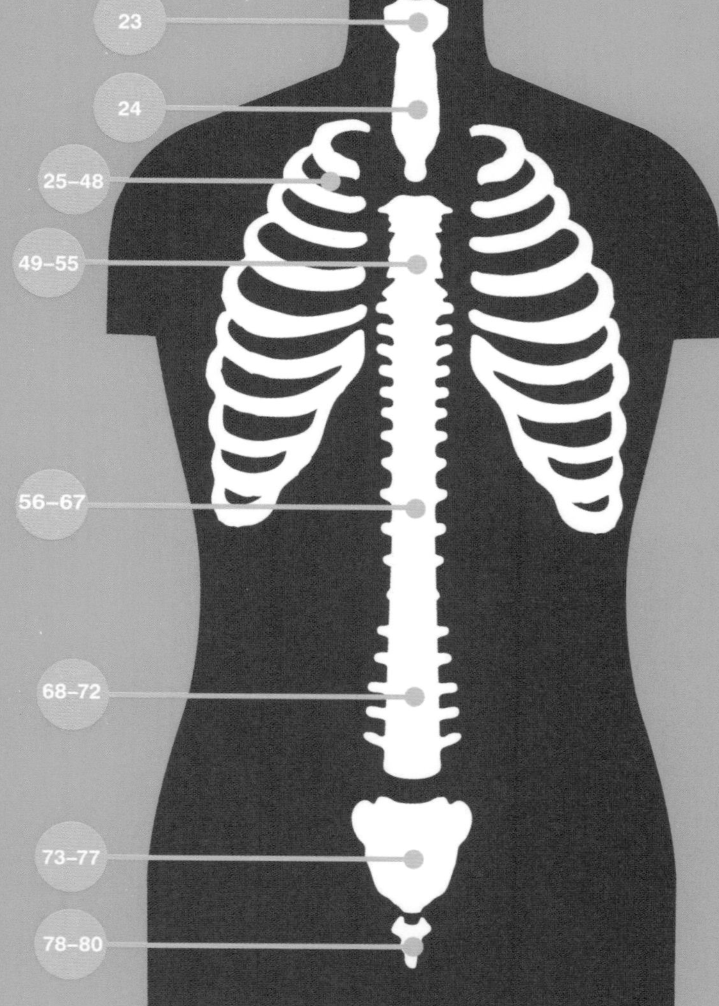

80 AXIAL SKELETON

Composed of four parts
Cranium, face, spinal column and the thorax

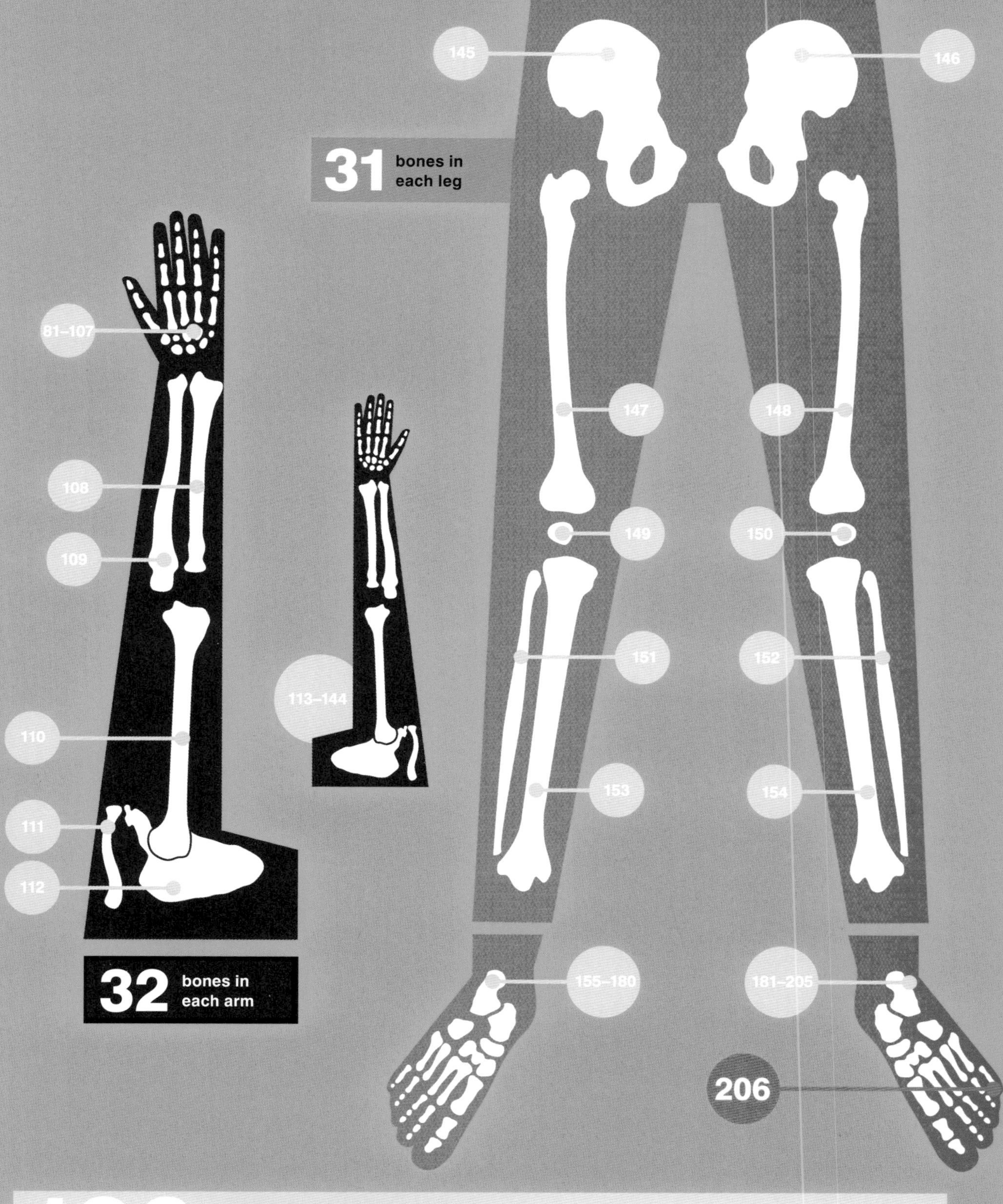

126 APPENDICULAR SKELETON

Composed of two parts
Arms and legs

DENTAL MATTERS

No part of the body is as hard as the enamel layer covering each tooth. Just under this layer is dentine, also tough and hard-wearing. And anchoring each tooth into its socket in the jawbone is the 'living glue' of cementum, another robust and resistant material. The whole package – in fact 32 packages, if all adult teeth develop and persist – facilitates almost a lifetime of bites, chews, gnashes and gnaws, and also grins and smiles.

ADULT TEETH

32: 8 Incisors, 4 Canines, 8 Premolars, 12 Molars

BABY TEETH

20: 8 Incisors, 4 Canines, 0 Premolars, 8 Molars

UPPER TEETH

Tooth	Eruption time (years)
First incisor	7–8
Second incisor	8–9
Canine	11–12
First premolar	10–11
Second premolar	11–12
First molar	6–7
Second molar	12–13
Third molar	17–21

Eruption times in years

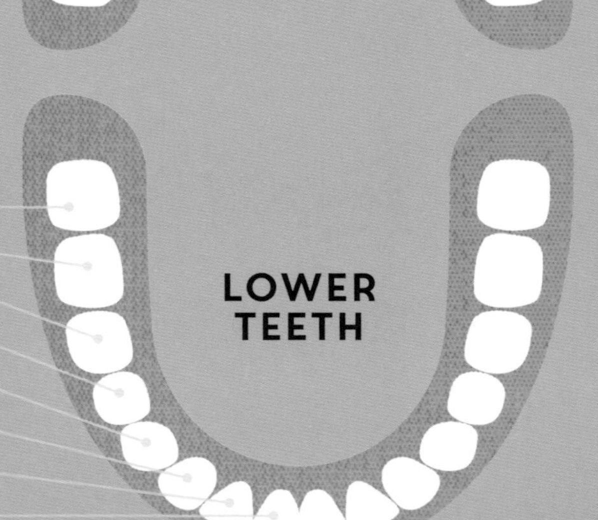

Tooth	Eruption time (years)
Third molar	17–21
Second molar	11–13
First molar	6–7
Second premolar	11–12
First premolar	10–11
Canine	9–10
Second incisor	7–8
First incisor	6–7

6–10
First incisor (lower jaw)

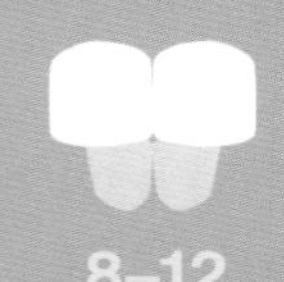

8–12
First incisor (upper jaw)

9–13
Second incisor

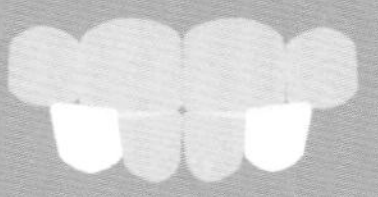

10–15
Second incisor

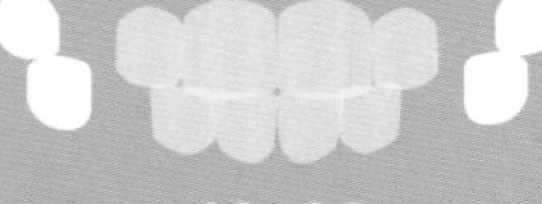

12–20
First molars

16–25
Canines

24–36
Second molars

Eruption times in months

HOW MANY ROOTS?

Incisors, canines, most premolars

Upper (maxillary) first premolars, lower (mandibular) molars

Upper (maxillary) molars

Wise teeth

Wisdom teeth are the four third molars, one at the rear of each side of each jaw. They usually erupt, if at all, when the individual becomes a 'wise' adult, at 17–21 years. But their appearance is variable. They may never really develop, or grow but not erupt, or erupt normally, or grow 'sideways' and press into or impact adjacent teeth.

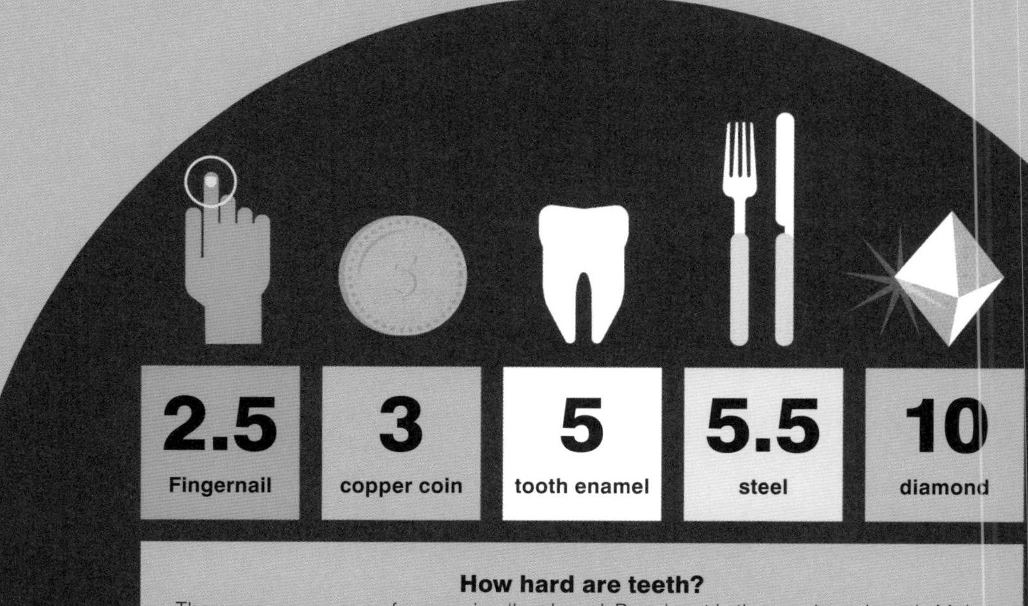

How hard are teeth?

There are many ways of measuring 'hardness'. Prominent is the rough-and-ready Mohs Scale used for minerals, based on what scratches what, with 10 standards as measures.

LOTS OF LENGTHS

About one-sixth of the body's weight is tubular organs. The blood, lymph, digestive and urinary systems are basically networks of fluid-containing pipes, tubes and tubules of diameters varying from greater than that of a thumb to one-tenth of a scalp hair. These various vessels are looped, folded and coiled with incredible intricacy and intimacy to fit within the human form. But uncoiled and unlooped, and joined end to end, their immense lengths become astonishingly apparent.

DIGESTIVE SYSTEM: mouth + throat + oesophagus + stomach + small intestine + ascending colon + transverse colon + descending colon + sigmoid colon + rectum + anus

9.5 m

URINARY SYSTEM

Nephron tubules (filtering units) in the kidneys.

50 km

The Grand Canyon 29km

Madrid

Paris

TOTAL LENGTH OF

2½ TIMES AROUND EARTH!

CARDIOVASCULAR SYSTEM

Capillaries **50,000**

Arterioles and venules **49,000**

Medium and large arteries and veins **1,000**

100,000 km

LYMPHATIC SYSTEM

Average number of lymph nodes for each area:
Abdomen **260** Neck **150** Groin **40** Axillary **40**

400–700

Berlin

Warsaw

Minsk

Moscow

NODES AND LYMPHATIC VESSELS (KM) 4,000

MUSCLE RECORDS

FASTEST CONTRACTION

Extra-oculars
Around sides and rear of eyeball. Rotate and swivel eyeball.

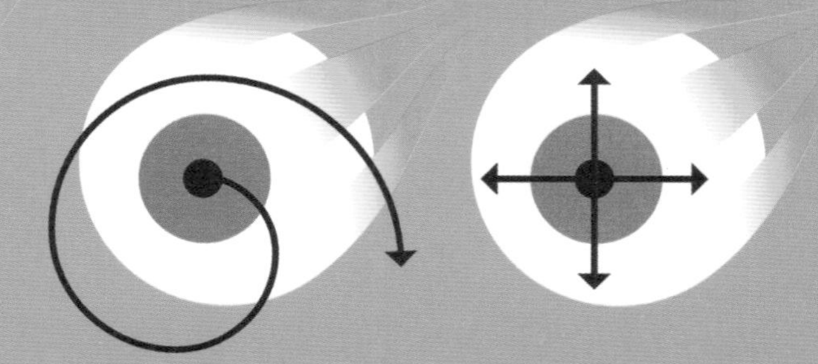

LONGEST

Sartorius
Crosses over front of thigh. Twists and raises thigh.

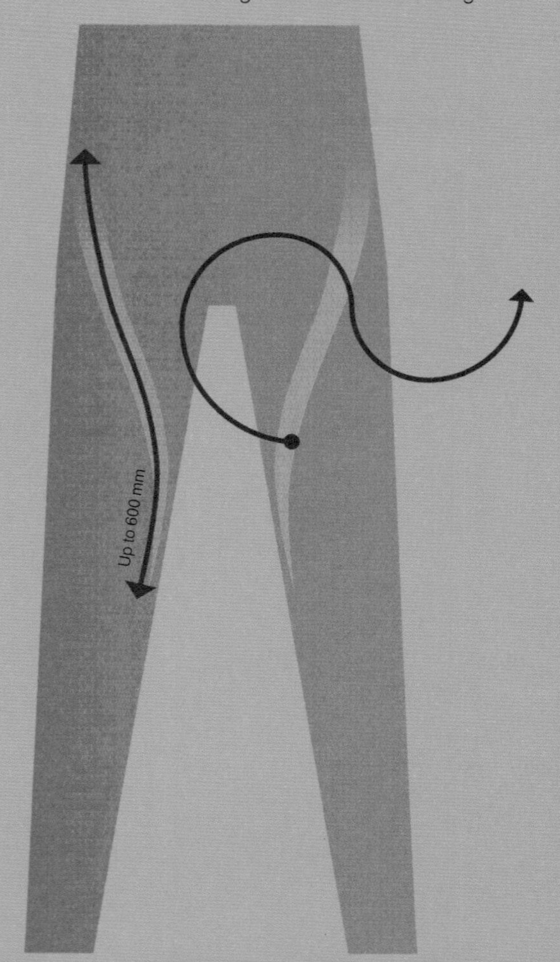

MEAN MUSCLES

About two-fifths of the body's weight is made up of muscle. There are more than 640 muscles cloaking virtually every part of the body, from the occipitofrontalis found in the forehead to the intrinsic plantar muscles of the sole of the foot.

One characteristic of muscles is their long and complicated names. These may derive from them being on the front (anterior, ventral) or back (posterior, dorsal), and so on, according to anatomical conventions. Or it may be the bone(s) to which they join, and sometimes the nerves they run alongside. Perhaps a nearby major organ. Maybe the motion they effect: flexors bend and extensors straighten. Another option is the muscle's shape: the deltoid in the shoulder is approximately triangular (like a river delta or the Greek letter delta). Indeed, for some unlucky muscles, almost all of these factors contribute to their elongated names.

BENDIEST

Superior lingualis
Upper surface of tongue (which is actually a complex of 12 muscles). Contributes to the tongue's vast range of movements.

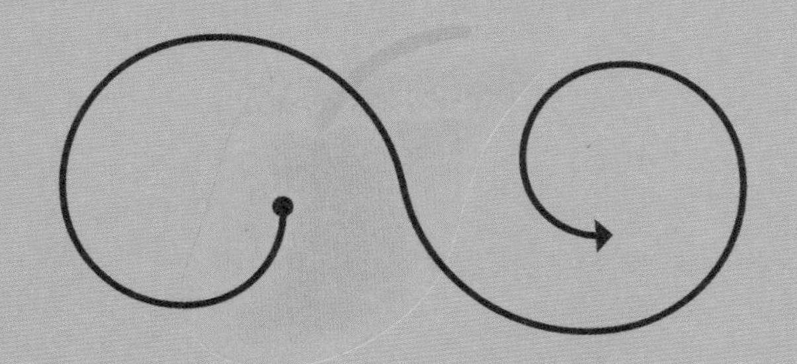

STRONGEST FOR SIZE

Masseter
Side of face and head, Biting and chewing.

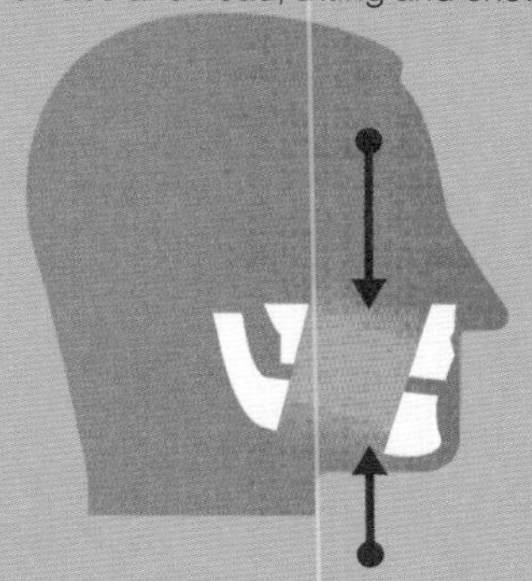

LONGEST MUSCLE NAME

levator labii superioris alaeque nasi

Lifts	the upper lip	and flares the nose's lower side

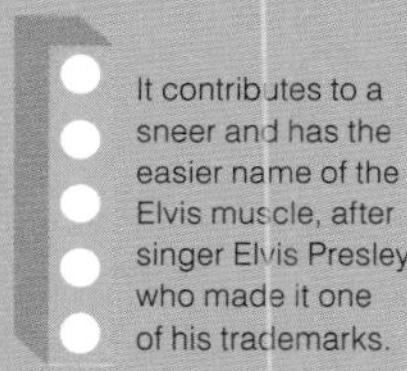

It contributes to a sneer and has the easier name of the Elvis muscle, after singer Elvis Presley who made it one of his trademarks.

SMALLEST

Stapedius
Inside inner ear. Dampens vibration from excessive noise.

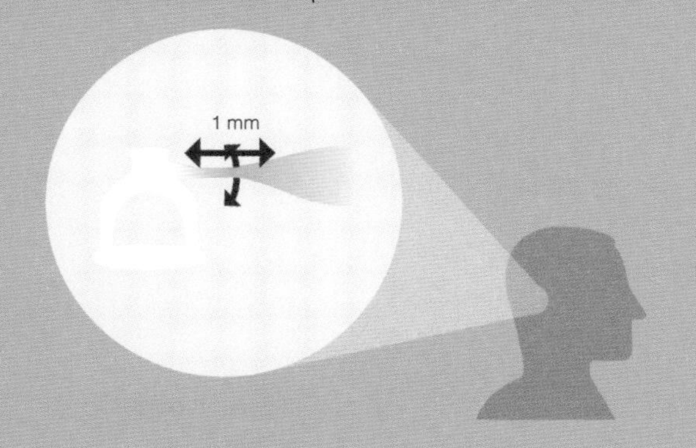

BULKIEST

Gluteus maximus
Forms most of buttock. Pulls thigh rearwards to walk, jump, run.

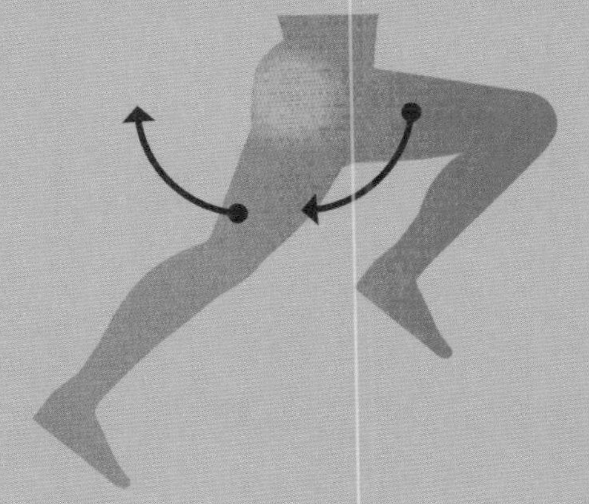

PULLING POWER

Living muscles are power-packed for their size and weight, but measuring human strength, power and work capacity is fraught with problems. A single muscle's contraction depends on its basic fitness (especially regular healthy usage or not), the speed of contraction and numbers of fibres involved (which depends on controlling nerve signals), whether the muscle is already partially contracted or fully relaxed, if it has just been pulling and so may be fatigued, and many other factors.

All together now
It's estimated that if all the body's muscles could contribute to one pulling force, they could lift 20 tonnes, about three African elephants.

Basic force
A muscle with cross-section area of one square centimetre exerts a maximum force of 40 newtons – enough to lift a 4-kilogram weight.

SOME COMPARISONS

Power output, W (watt, mouse) or kW (1,000 watts, other) / Power-to-weight ratio, W/kg (watts per kilogram)

0.2 / 5

1–1.5 / 3.5

10 / 20

100 / 60

INSIDE THE INSIDE OF THE INSIDE OF INSIDE A MUSCLE

Muscle e.g. biceps brachii (upper arm)
Relaxed length: 250 mm
Maximum diameter, contracted: 65 sq cm

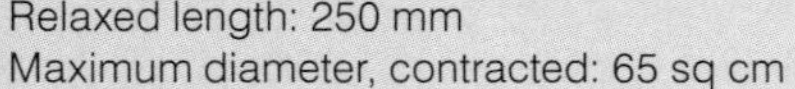

In theory
A fit biceps brachii arm muscle with a maximum cross-section area of 65 square centimetres could in theory raise 260 kilograms – equivalent to three or four adults.

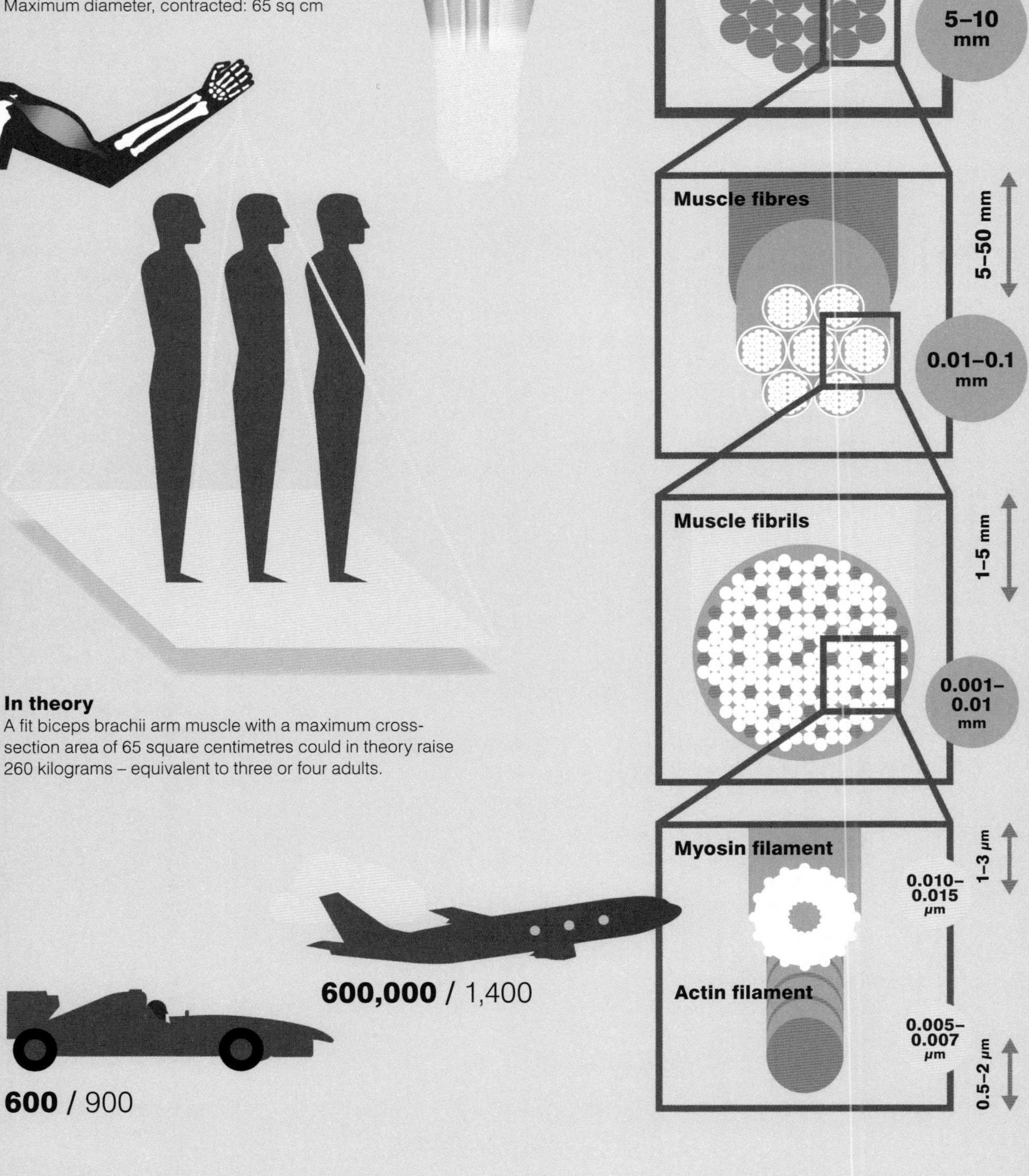

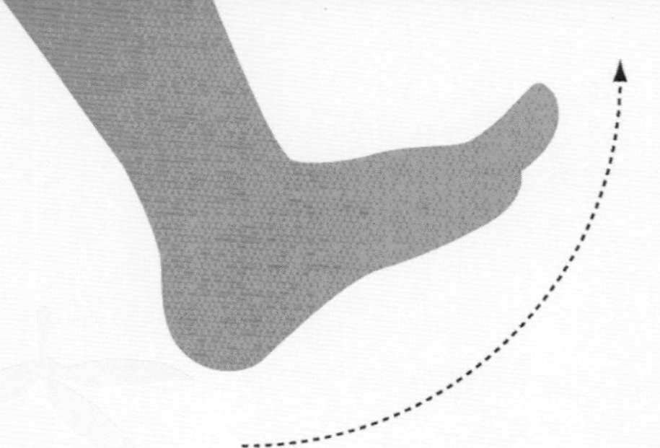

SLIPPERY OR WHAT?

Coefficient of kinetic friction, lubricated[1], Adjacent materials

0.003	Cartilage + synovial fluid
0.005	Ice skate blade + ice
0.02	Ice + ice
0.02	BAM + BAM[2]
0.04	PTFE + PTFE[3]
0.05	Ski + snow
0.2	Steel + brass
0.5	Steel + aluminium
0.8	Rubber + concrete

1 Resistance to sliding when already moving

2 Boron– aluminium–magnesium, one of the slipperiest man-made solids

3 Polytetrafluoroethylene, brand names include Teflon

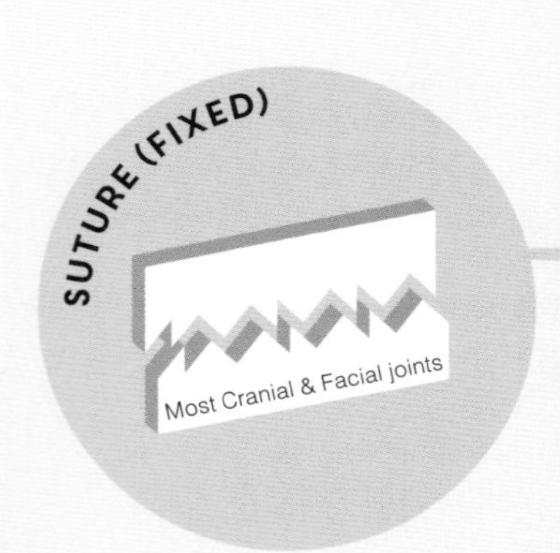

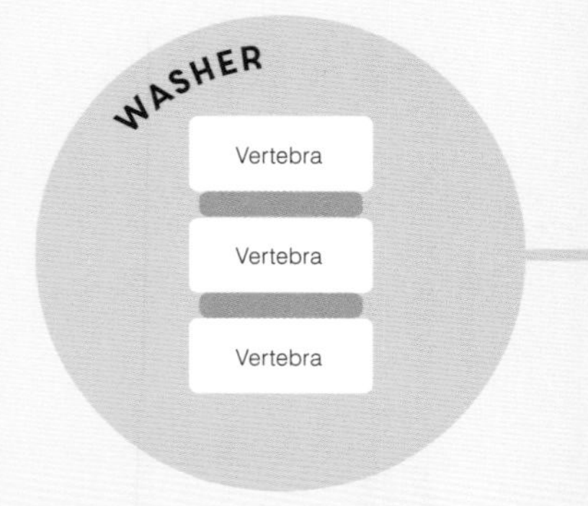

JOINED AT JOINTS

The human skeleton has between 170 and 400 joints, depending on how they are defined – whether three bones coming together, all in at least some kind of contact with each other, is one, two or three joints. These punishingly physical parts work so well for so long because the bone ends are covered with slightly soft-cushioned, smooth-gliding cartilage and lubricated by ultra-slick synovial fluid. Also, a tough bag-like capsule encases the joint, and stretchy ligaments link the bones to allow motion yet prevent dislocation, when bone ends separate – a pain that, once felt, is seldom forgotten.

CONDYLOID

140

TOES

Indicates a typical range in degrees for the joint's flexibility in young adults

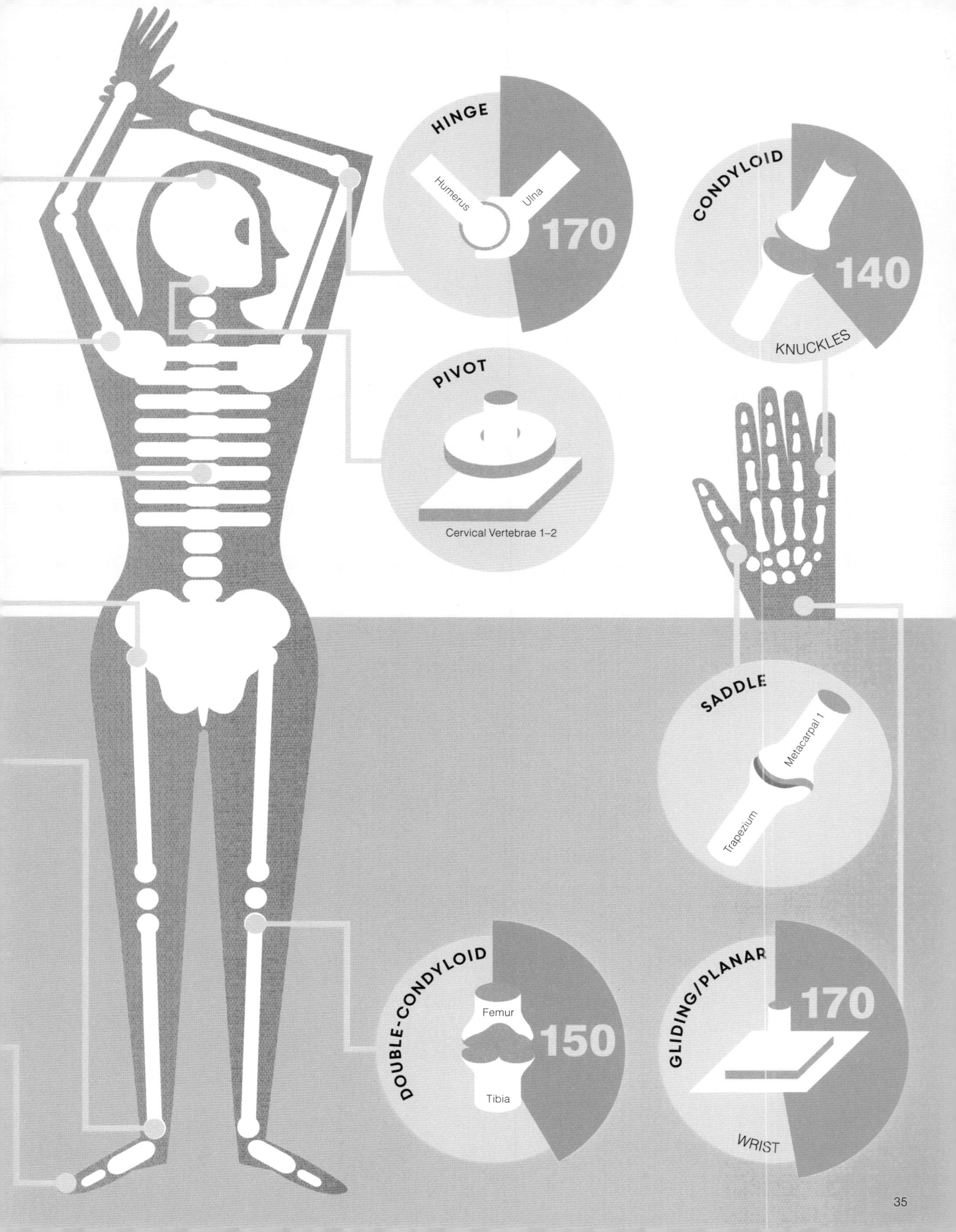
HINGE
Humerus
Ulna
170
CONDYLOID
140
KNUCKLES
PIVOT
Cervical Vertebrae 1–2
SADDLE
Metacarpal 1
Trapezium
DOUBLE-CONDYLOID
Femur
150
Tibia
GLIDING/PLANAR
170
WRIST

BREATH OF LIFE

Take a deep breath. Breathe in a bit more. And more, keep going…
Even the biggest inspiration is unlikely to fill the lungs. The aim of the breathing process and bodily respiration (as opposed to cellular respiration) is to take fresh air into the lungs. From here oxygen passes into the blood stream and on to the next system, cardiovascular, for bodywide distribution. A secondary goal of respiration is to expel the waste product carbon dioxide (made by cellular respiration) which, if it rises just 10–20 % above normal, can precipitate gasping, dizziness, even unconsciousness. A third useful effect of breathing is speech and other vocalizations. So the airways, lungs and chest muscles keep on breathing, in and out, around 8–10 million times each year.

If you sneezed your very hardest, the air speed out of your nose would be 20 m per second or 72 km/h

810 m

Total air breathed in a lifetime (litres)

280,000,000

GASES IN %

78 Nitrogen

Oxygen **21**

Others less than **1**

Carbon dioxide **0.3**

Water vapour atmospheric (varies)

400–600
million alveoli (tiny air sacs)

2,500
km bronchi, bronchiole airways

1,000
km capillaries (tiny blood vessels)

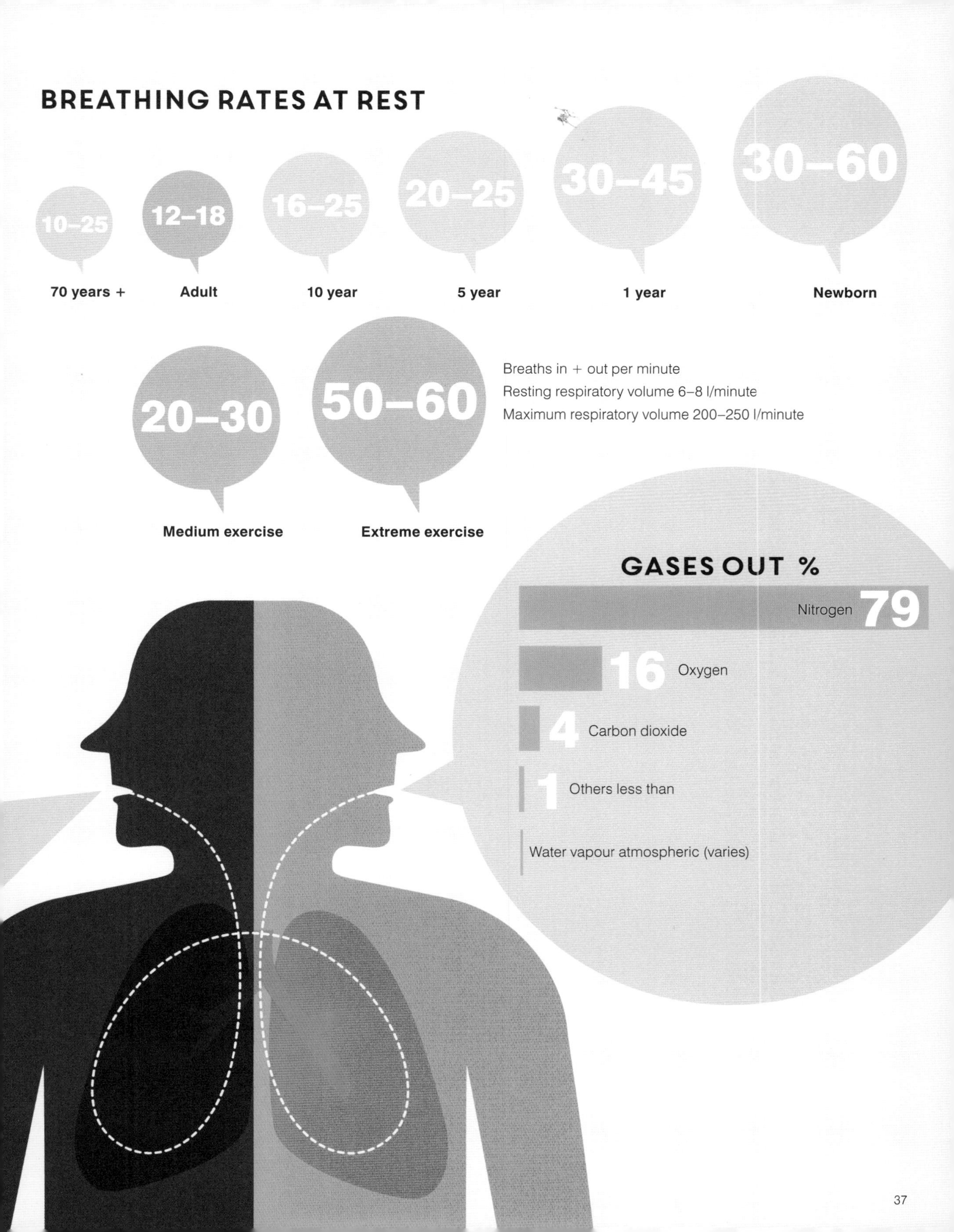
BREATHING RATES AT REST
10–25
70 years +
12–18
Adult
16–25
10 year
20–25
5 year
30–45
1 year
30–60
Newborn
20–30
Medium exercise
50–60
Extreme exercise
Breaths in + out per minute
Resting respiratory volume 6–8 l/minute
Maximum respiratory volume 200–250 l/minute
GASES OUT %
Nitrogen 79
16 Oxygen
4 Carbon dioxide
1 Others less than
Water vapour atmospheric (varies)

VITAL BEAT

The heart – a seemingly simple double-pumping bag of muscle – beats 3,000 million times or more through a life. When it stops, so does that life (unless critical medical help is at hand). In fact, the heart and its blood system are incredibly complex. The heart itself, even without its body, has an innate or intrinsic contraction rate of 60–100 beats per minute, due to its own set of natural pacemakers. Influences from the body – chiefly signals along the vagus nerve from the brain, and hormones such as adrenaline (epinephrine) – alter this speed, and also the volume and force of each beat, to match the body's immense range of needs.

Carotid
Neck

Pulse
A surge of high-pressure blood with each beat spreads out along the arterial vessels.

Felt most easily where an artery is just under the skin and has stiff tissue behind it. Usually at the radial artery in the wrist, just below the thumb mound.

Brachial
inside of elbow

Radial
wrist

Femoral
groin

Popliteal
rear of knee

Heart rates at rest with age (beats/minute)

120	90	80
Newborn	1 year	10 year
60-80	**40-60**	**58-80**
Adult	Sportsperson	70+ years

Energy
Each day the heart muscles produce enough motion energy to power a truck for 30 km.

At rest
A heart would take 30 minutes to fill a bathtub with blood, and an Olympic swimming pool, 5 years.

Dorsal pedal
top of foot

Posterior tibial
ankle

PHYSICAL HEART

Size approximately that of its owner's clenched fist

350

Average mass in grams

UNDER PRESSURE

Almost every body part* – every individual cell – relies on flowing blood to bring oxygen and nutrients, and to wash away carbon dioxide and other wastes. Flow is produced by the heartbeat, in two main phases. During diastole ('dye-ass-toll-ee') the organ's muscular walls relax, and it enlarges as blood oozes in at low pressure from the veins – wide, floppy, thin-walled tubes that return blood from the smallest vessels, capillaries, to the heart. Just half a second later comes systole ('siss-toll-ee') when heart muscles tense and contract, forcing blood to rush out away from the heart at high pressure though the thick-walled, muscular arteries, which divide eventually to form capillaries. The pressures involved here are greatest of any body system and make the vessels bulge in travelling waves that spread through their branching network. (*Among the few body parts that do not have a direct blood supply are the eye's cornea and lens; if they did our world view would be obscured by a reddish mesh.)

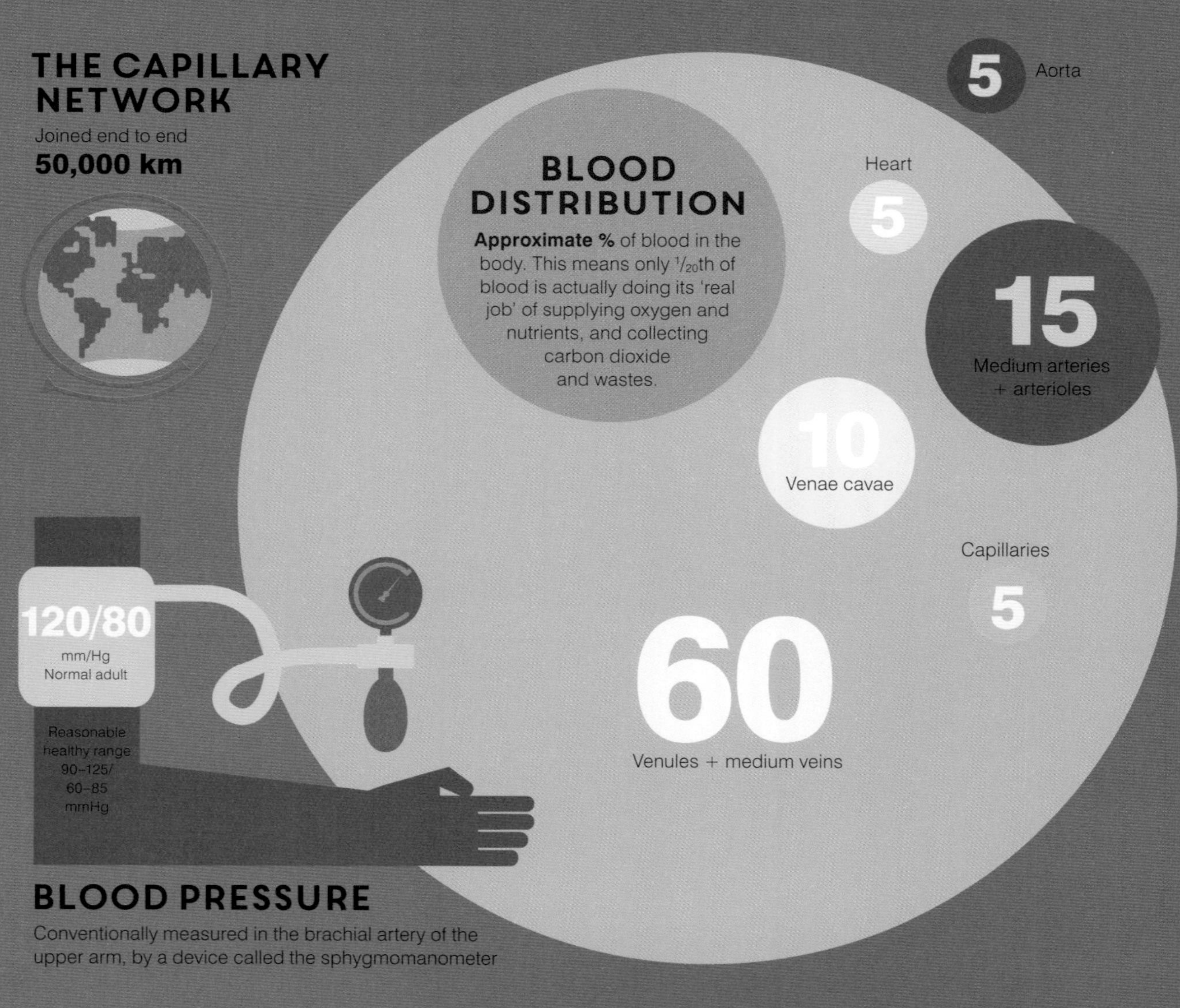

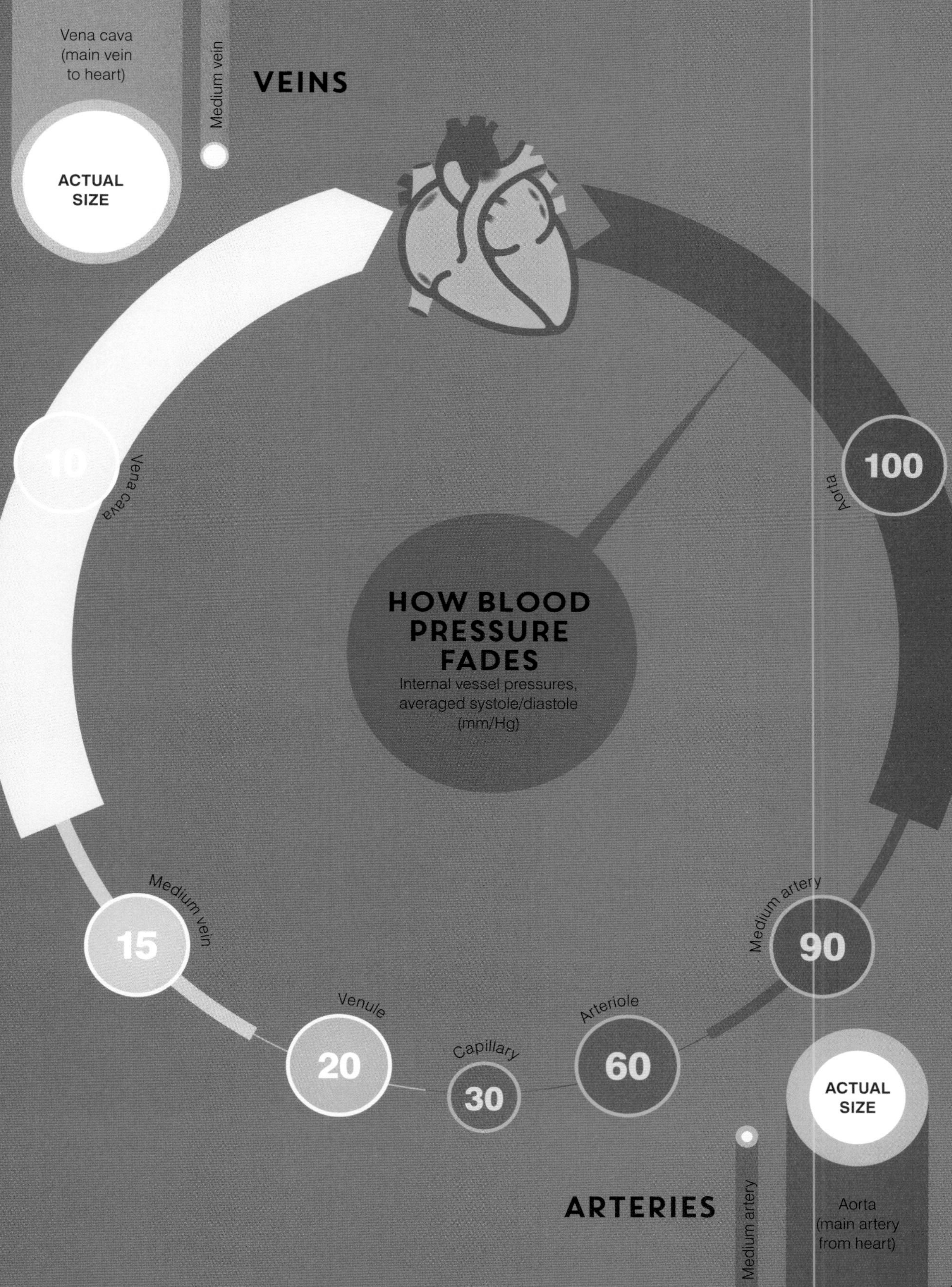
Vena cava
(main vein
to heart)
ACTUAL
SIZE
Medium vein
VEINS
HOW BLOOD
PRESSURE
FADES
Internal vessel pressures,
averaged systole/diastole
(mm/Hg)
10
Vena cava
100
Aorta
Medium vein
15
Venule
20
Capillary
30
Arteriole
60
Medium artery
90
ACTUAL
SIZE
ARTERIES
Medium artery
Aorta
(main artery
from heart)

WHAT MAKES A CHAMPION?

The body recipe for a champion athlete is complex, with many different factors at work. They include opportunities to train; the qualities of the coach, nutritionist, physiologist and other experts; and the equipment, location and other facilities. A vast amount is in the mind: self-motivation, thorough application, and the will to win; support from family and friends plays an important role here. But perhaps most important is a person's contributing genes: how you are built may equip you better for one sport over another.

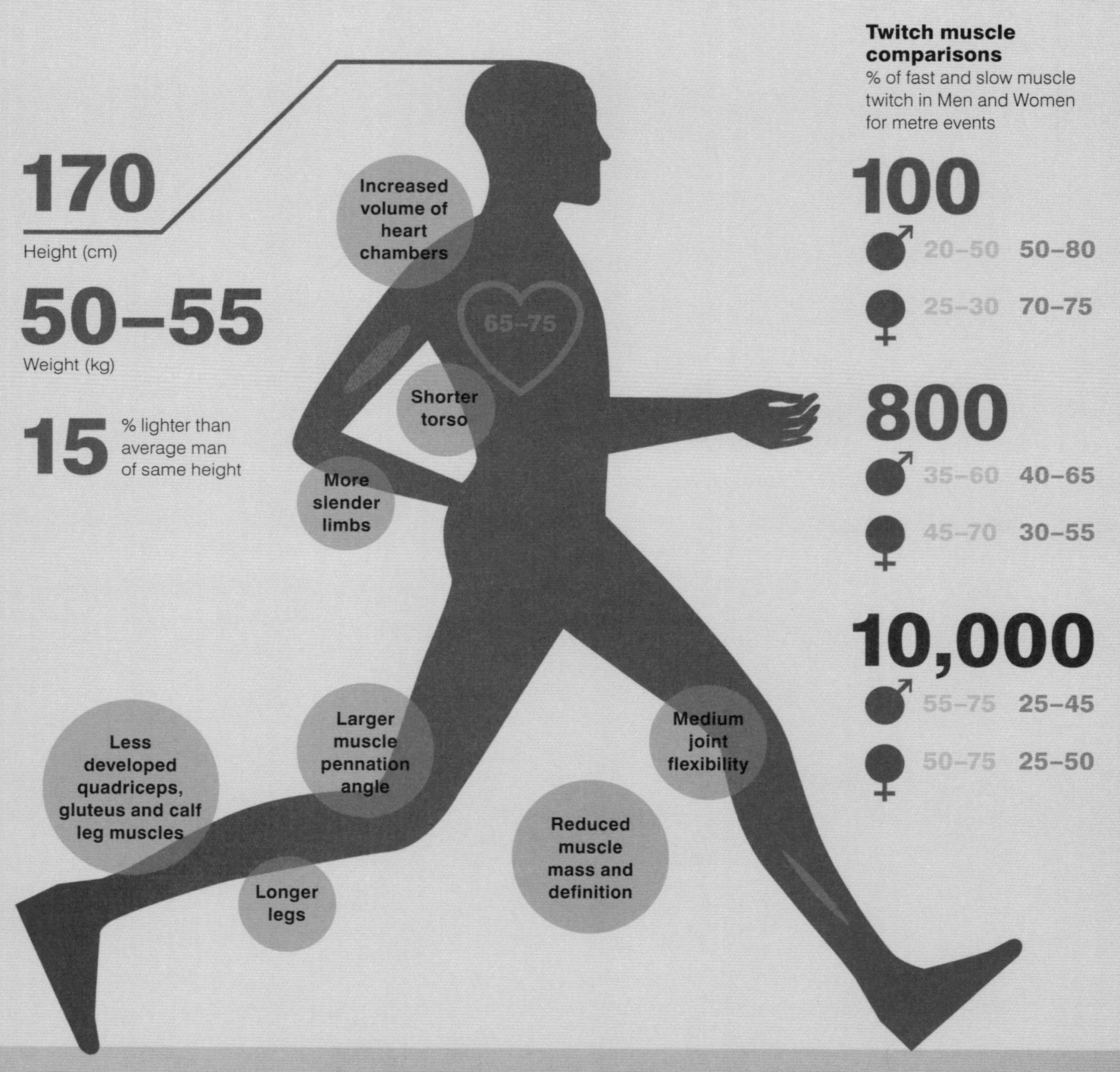

ENDURANCE RUNNER

Very lean (almost zero fat)

TWITCHY MUSCLES

Most muscles in most people have two kinds of fibres. **Slow twitch** (type I) contract more slowly, generate less force, but can work for longer before fatigue. **Fast twitch** (type II) contract rapidly, generate short bursts of strength or speed, but fatigue quickly. Different kinds of training maximize growth and strength of existing fibres, to change their relative contributions to movements. Less intense exercise promotes slow-twitch fibre development, more intensity encourages fast-twitch fibres. The balance of fast- and slow-twitch fibres is determined by genetics. A 'strong' version of the gene ACTN3 enhances the proportion of fast-twitch fibres.

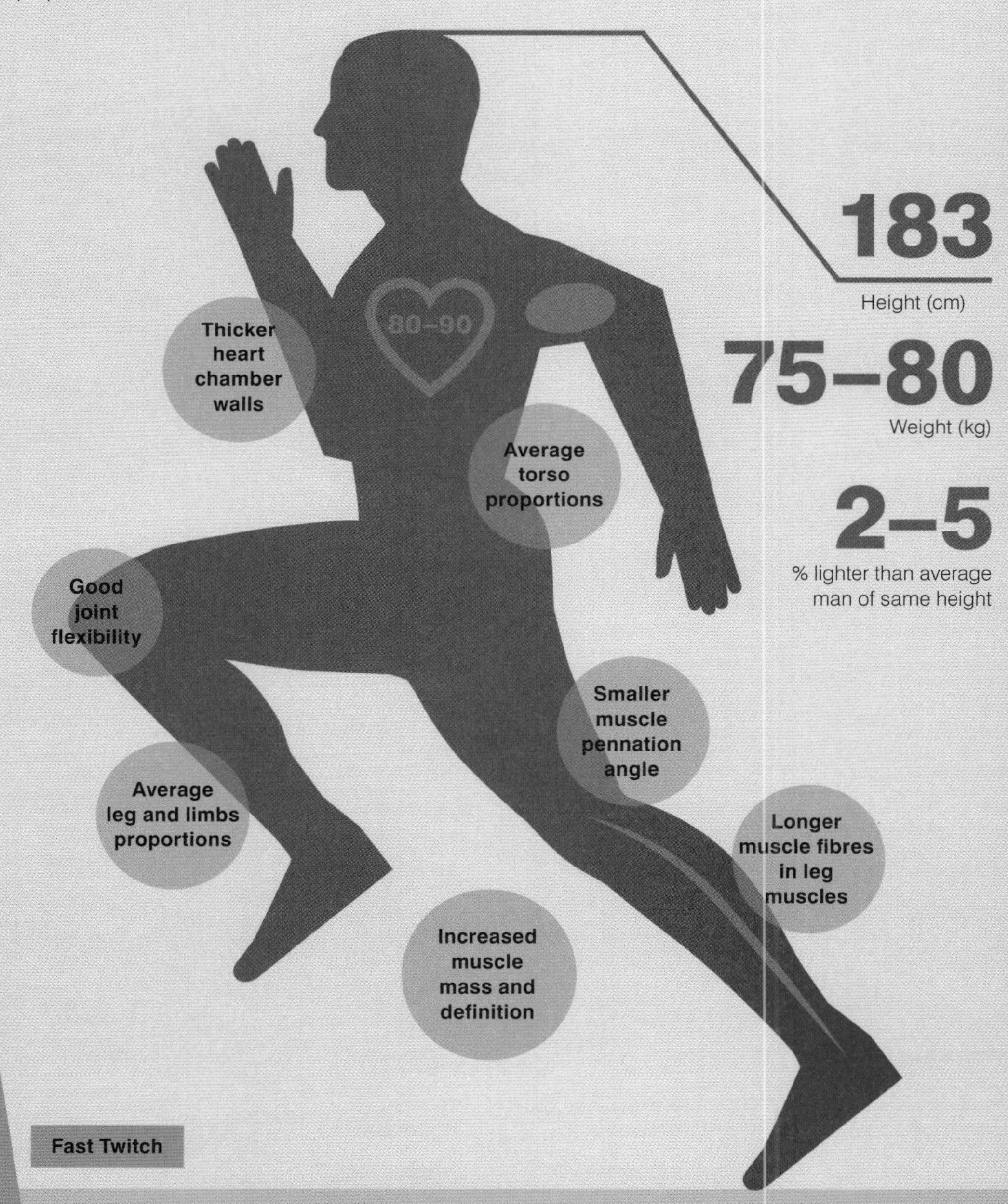

SPRINT RUNNER

Lean (minimal fat)

FASTER, HIGHER, STRONGER

In 1924 *Citius*, *Altius*, *Fortius* ('Faster, Higher, Stronger') became the official motto of the modern Olympic Games, which had kicked off in 1896. The saying celebrates how, since first staged in Ancient Greece, the human body's athleticism and other skills are taken to the limit and receive global recognition. The Games' 20-plus sports are a world benchmark for the body's physical powers. Since that time, the body's Olympian triumphs have advanced in speed, height and strength. Many factors are at work in this, however. There has been steady progress in diet, hygiene and general health as well as improvements in specialized skills, training, coaching and equipment. The late 1930s and early 1940s were interrupted by war. The 1950s–60s saw considerable suspicions about steroid and other drug misuse. And there are occasional quantum shifts in a sport's technique, such as the 'Fosbury flop' in high jump introduced in the 1968 Olympics. The Olympics remain the standard by which we measure what the body, taken to the limit, can achieve.

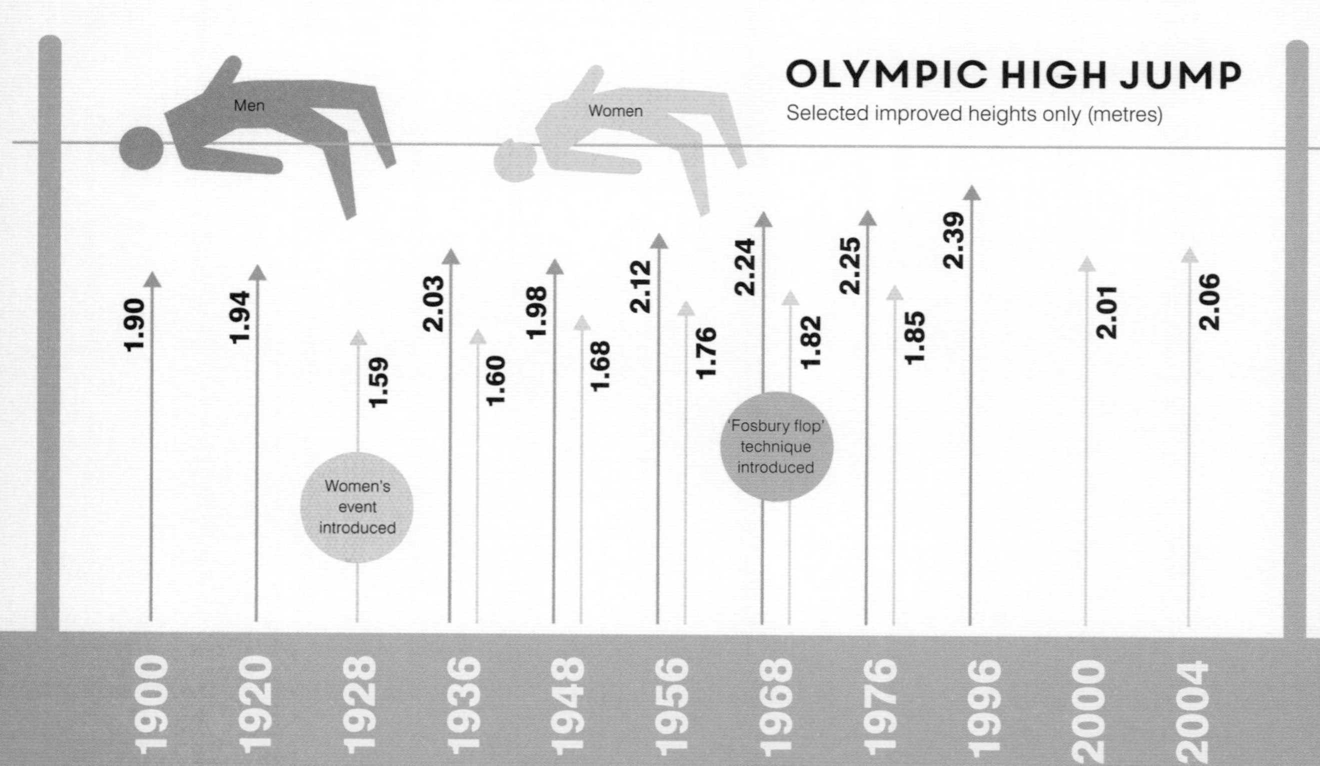

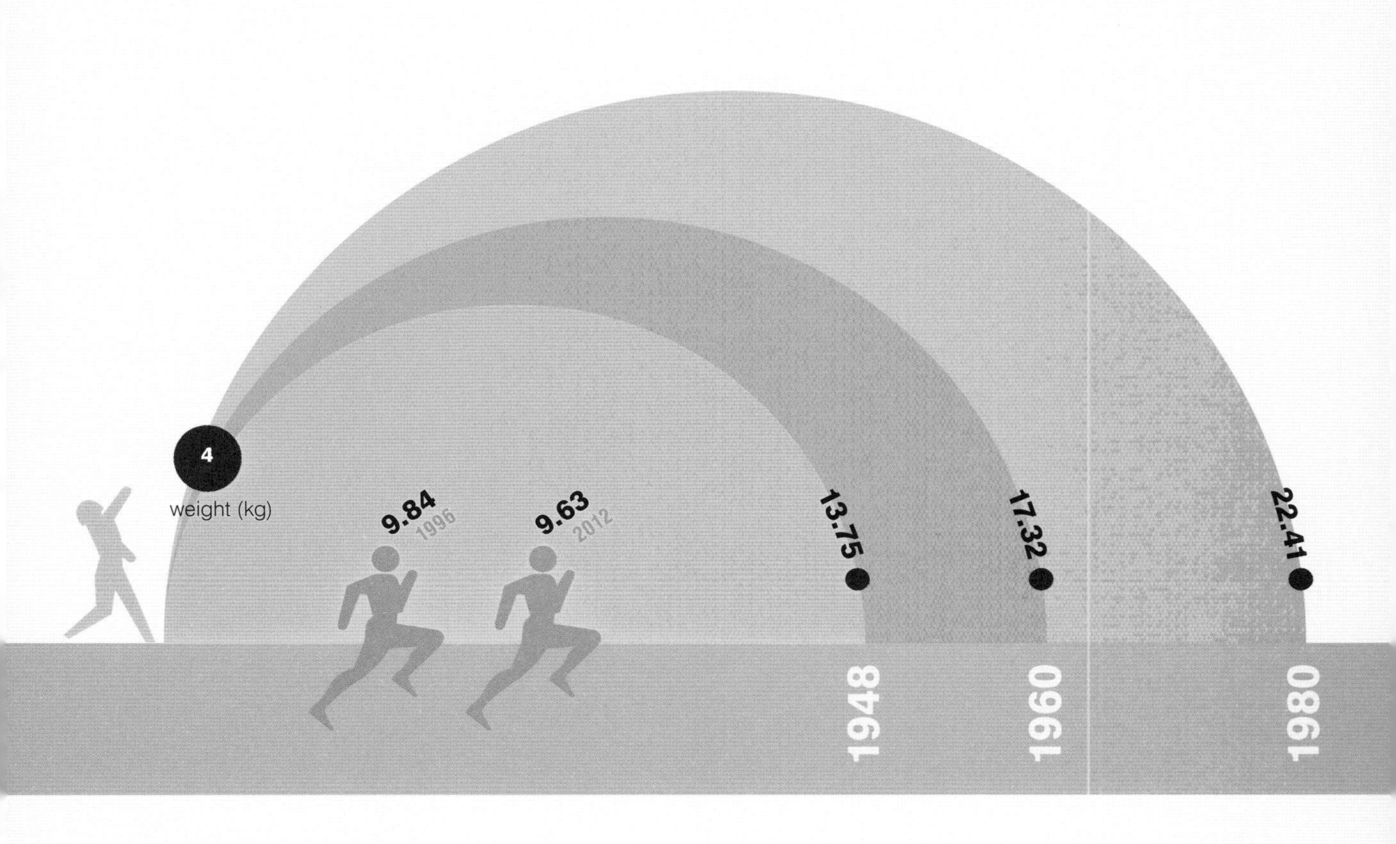

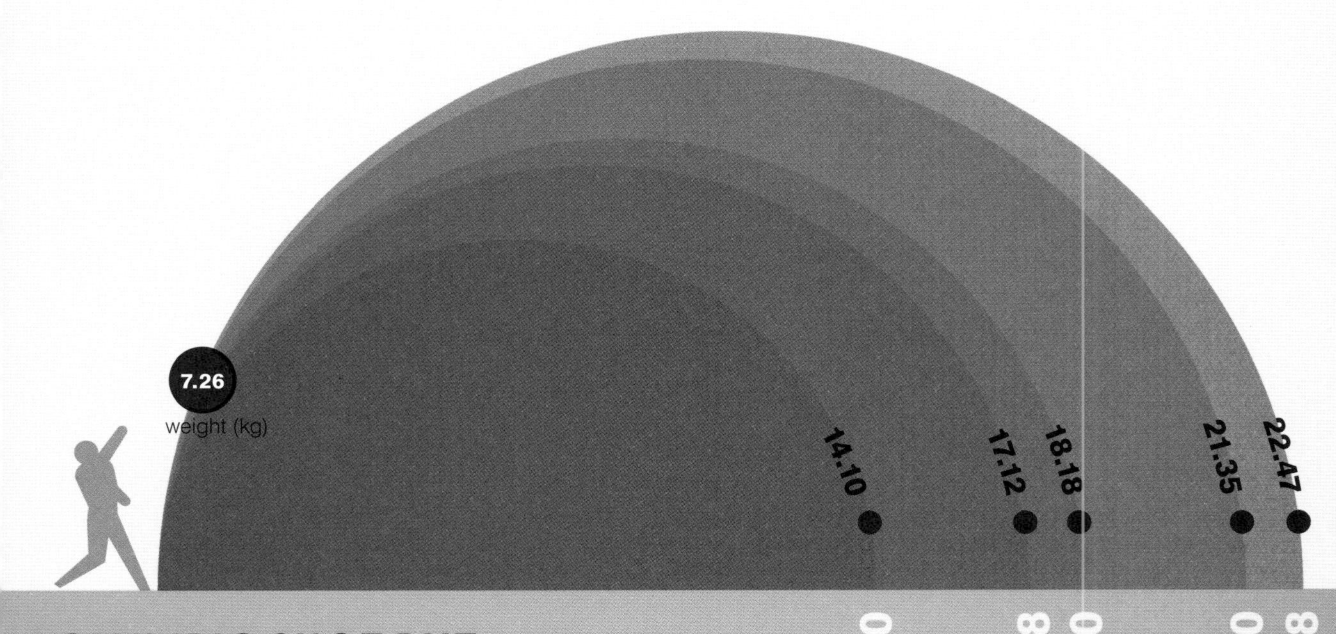

OLYMPIC SHOT PUT

Selected improved distance only (metres)

CHEMICAL BODY

THE CHEMICAL FACTORY

Everything is made of atoms. The human body is no exception and estimates abound as to its proportions of various pure chemical substances. How are these inventories compiled? One way is to list elements by percentage mass (weight); this favours heavier elements such as iron, whose atoms are almost 56 times heavier than those of the lightest element, hydrogen. Another method is to itemize by numbers of atoms; since water (H_2O) makes up 60% of most bodies, this approach promotes its two elements, hydrogen and oxygen. So hydrogen comprises 9–10% of the body by mass, but 65–70% by numbers of atoms.

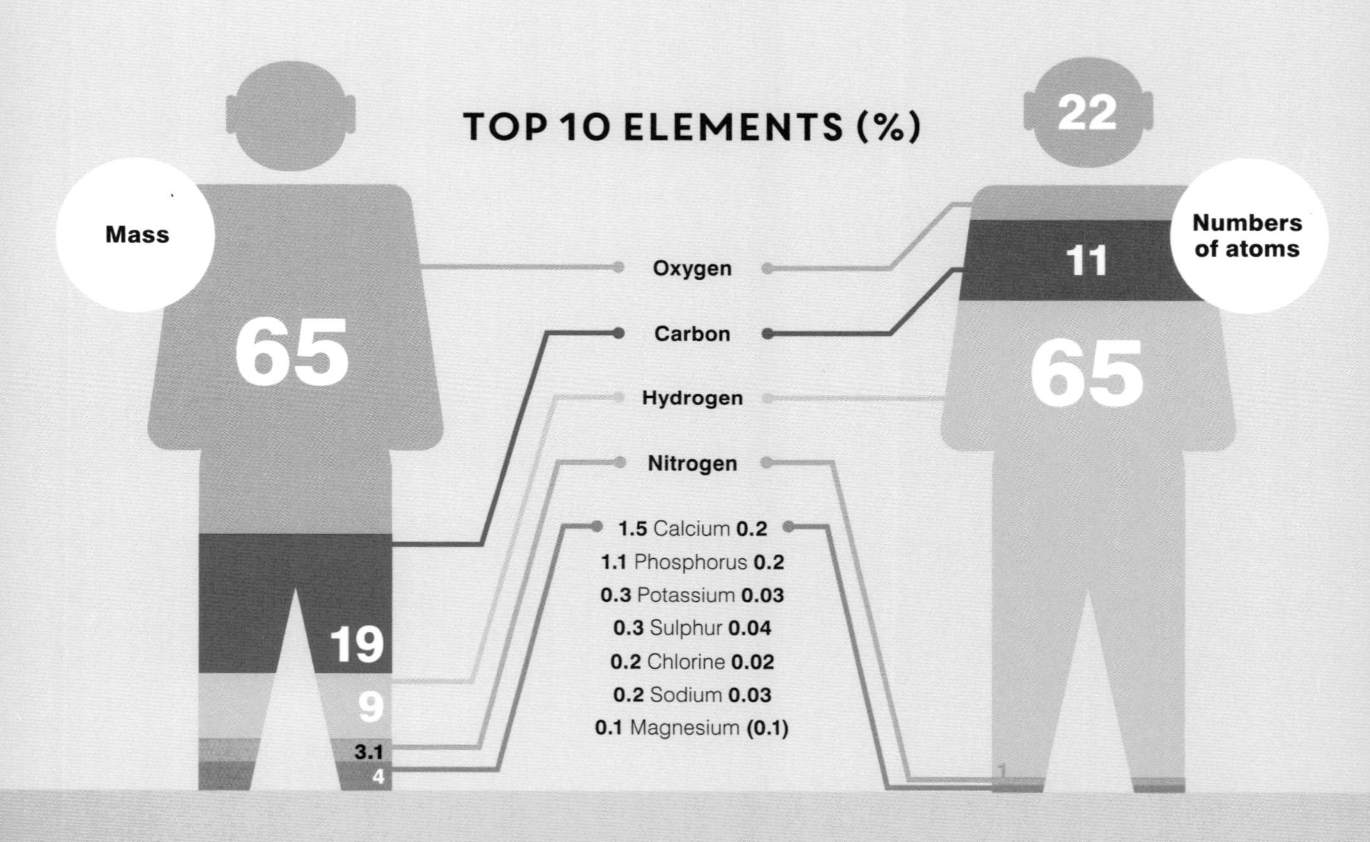

A 70 KG BODY HAS ENOUGH...

TRACE MINERAL ELEMENTS AT LESS THAN 0.1%

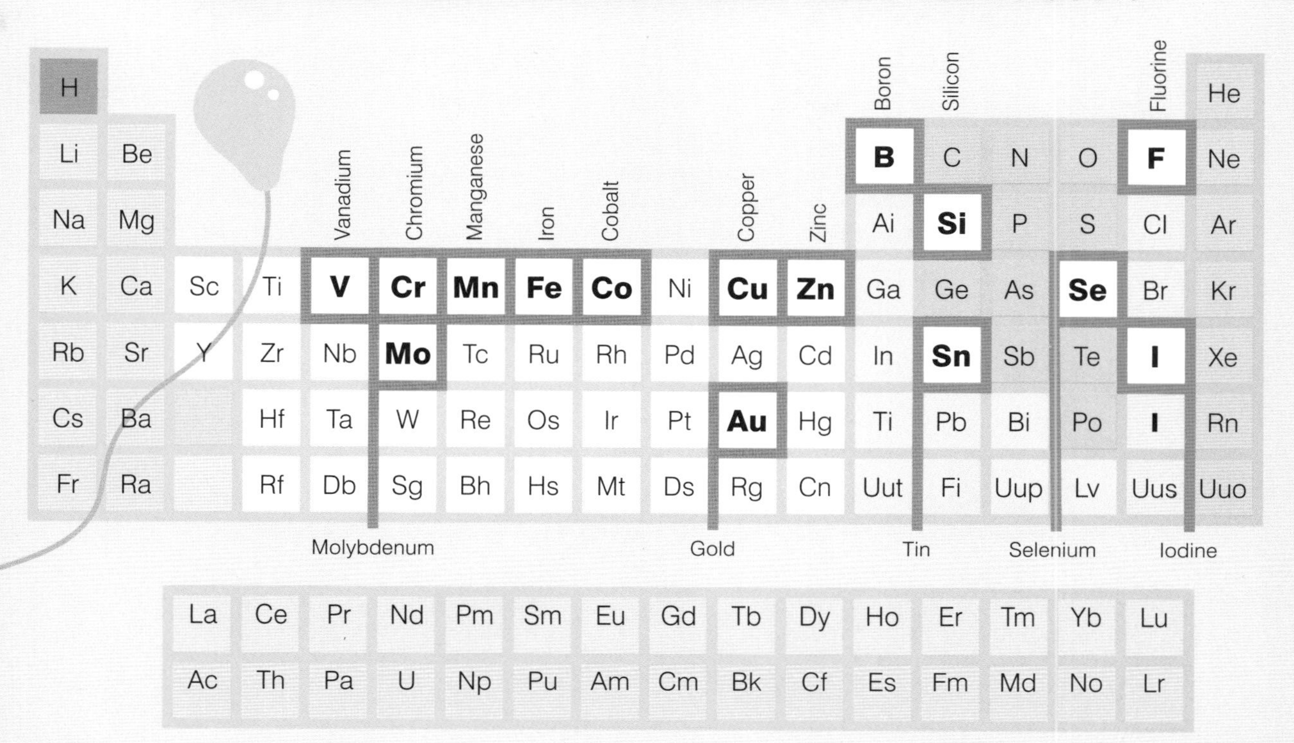

Mineral wealth?
All the elements of one body extracted and sold through global trading markets in would fetch around

£3,000

There's gold in them thar' bodies!
A human contains about 0.2 mg of gold, making a cube with 0.2 mm sides.

0.000,2 g

H **Hydrogen for...**

5,000
helium party balloons (6 kg)

C **Carbon for...**

10,000
graphite carbon pencil leads (13 kg)

P **Phosphorus for...**

20,000
match heads (800 g)

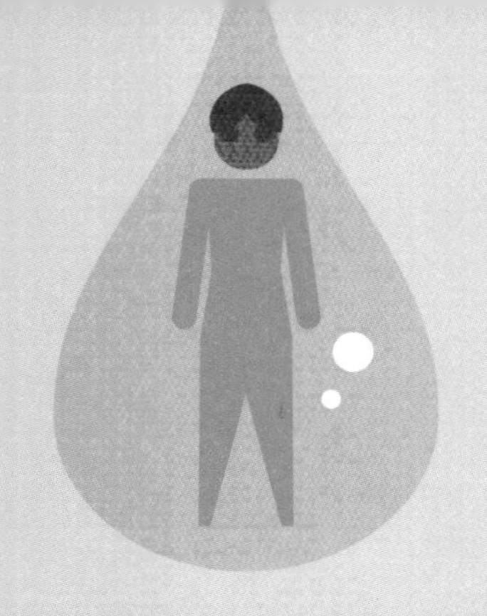

THE WET BODY

Human bodies are mostly water. The broad-spectrum, averaged proportion is two-thirds, which naturally varies according to conditions and circumstances. A higher proportion of body fat, for example, reduces the overall percentage, because fatty tissue contains much less water than other body tissues, including bone. Even so, the body has a lot of water – more than 45 litres for a 70 kg person, enough for a quick shower. With the water from three bodies, you could wallow in a reasonably large bathtub.

It's not possible to keep the water in the body, in the body. Water must leave to carry away dissolved and potentially harmful wastes, chiefly in urine. Around three litres daily generally suffices for this turnover. But it's more in hot conditions, when active, and when imbibing substances such as alcohol.

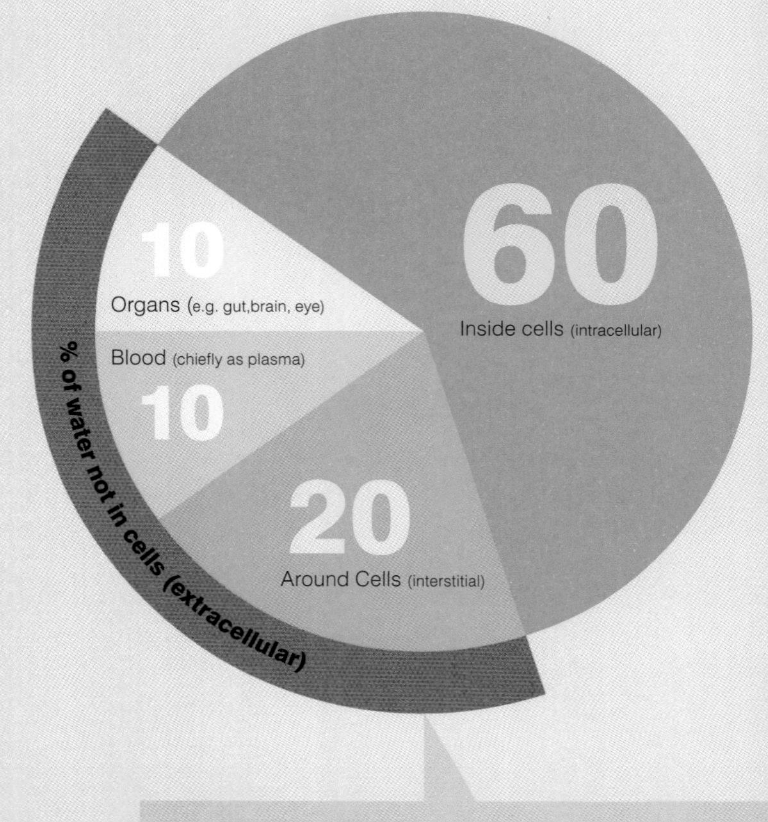

Where water is found
Biologists talk of 'compartments' of water. Not neat cupboards or chambers inside the body, rather estimated accumulated totals of water in and between and around the millions of cells, hundreds of tissues and dozens of organs.

Average water (by mass) with age in %

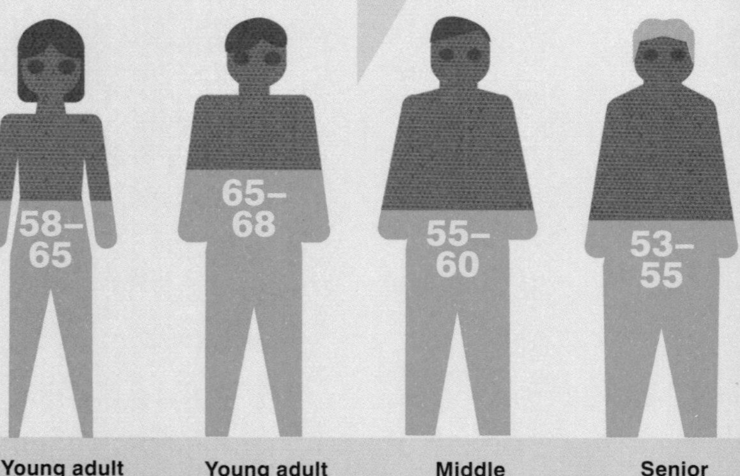

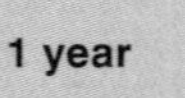

Newborn	1 year	Young adult female	Young adult male	Middle age	Senior (70+)
75	65	58–65	65–68	55–60	53–55

TURNOVER OF WATER PER DAY (mls) 2,700

750 FOOD
300 METABOLIC WATER[1]
1,650 DRINKS

200 FAECES
1,700 URINE
800 SKIN, LUNGS[2]

2,700

Water in organs and tissues

By mass in %. Includes their internal fluids e.g. blood, urine

Organ/tissue	%
Lungs	85
Blood	85
Kidneys	80
Muscles	75
Brain	75
Spleen	75
Heart	75
Digestive	70
Liver	70
Skin	65
Bones	25
Fat	10

1 When sugars and similar carbohydrates are broken down to release their energy, one natural byproduct of the chemical process is water. This contributes to the body's water income. $C_6H_{12}O_6 + 6CO_2 > 6CO_2 + 6H_2O$ + energy, or in words, sugar + oxygen > carbon dioxide + water + energy

2 Tiny amounts of water ooze from the skin in almost all conditions, known as 'insensible' sweat. Also, the air that is breathed out is almost saturated with water vapour evaporated from the moist linings of the lungs and airways.

MICRONUTRIENTS

The body needs many nutrients in quantities far less than the major macronutrients of carbohydrates, fats, proteins and dietary fibre. Most 'micros' are vitamins and minerals. Vitamins are organic substances that are needed for the body to run smoothly. Most must come ready-made in the diet since the human body cannot manufacture them itself in sufficient quantities. Minerals are simple chemical substances, for example, metals like sodium, iron, calcium, and manganese, and non-metals or salt-formers such as chloride, fluoride and iodide.

DAILY INTAKES[1] IN MILLIGRAMS[2]

3,000 Chloride[3] — Salt

900 Sulphur[4] — Eggs

800 Phosphorus — Pumpkin seeds

200 Potassium — Sweet potato

300 Magnesium — Spinach

MAJOR MINERALS

The body requires these major minerals in amounts of at least 100 milligrams (0.1 grams) per day.

2,000 Sodium — Salt

VITAMINS

Most vitamins are required in very small amounts, in some cases a few millionths of a gram.

75–90 C ascorbic acid

15 B3 niacin

B5 pantothenic acid **5**

20 E tocopherol

A retinol group **0.7–0.9**

1.5–1.7 B6 pyridoxine

B2 riboflavin **1–1.3**

1–1.2 B1 thiamin

Some vitmains are required in even smaller dosages

400–600 μg[5] B9/Bc/M folate, folic acid

90–120 μg K Phylloquinone, metaquinones

30 μg B7 Biotin

10–15 μg D cholecalciferol

2–2.5 μg B12 cobalamine

Relative scale of vitamins in relation to minerals.

90

18

18 Iron

Fluoride **4**

TRACE MINERALS

This is a far from complete list, which would continue for several dozen pages of this book.

2 Manganese

2 Copper

15 Zinc

Molybdenum

Iodide

Selenium

Chromium

Milk

1,000 Calcium

1 RDI Reference or Recommended Daily Intake; many similar categories exist such as RDA Recommended Daily Allowance, AI Adequate Intake

2 mg milligrams (0.001 or one-thousandth of a gram) unless stated

3 As sodium chloride (table or common salt)

4 Sulphur has no official RDI; quantities are based on average healthy intake

5 μg micrograms (0.000,001 or one-millionth of a gram; 0.001 or one-thousandth of a milligram)

MACRONUTRIENTS

Recommended macronutrients in grams for a standard daily food intake that supplies energy of 8,700 kJ (kilojules) or 2,100 Calories (kcal).

300–310
Carbohydrates

90
Glucose and other sugars

20–25
Saturated fatty acids

0.3 Cholesterol

65–70
Total fats

20–25
Dietary fibre

45–55
Total proteins

YOUR ORGANS AND ENERGY (%)

Main energy users for a fairly active individual.

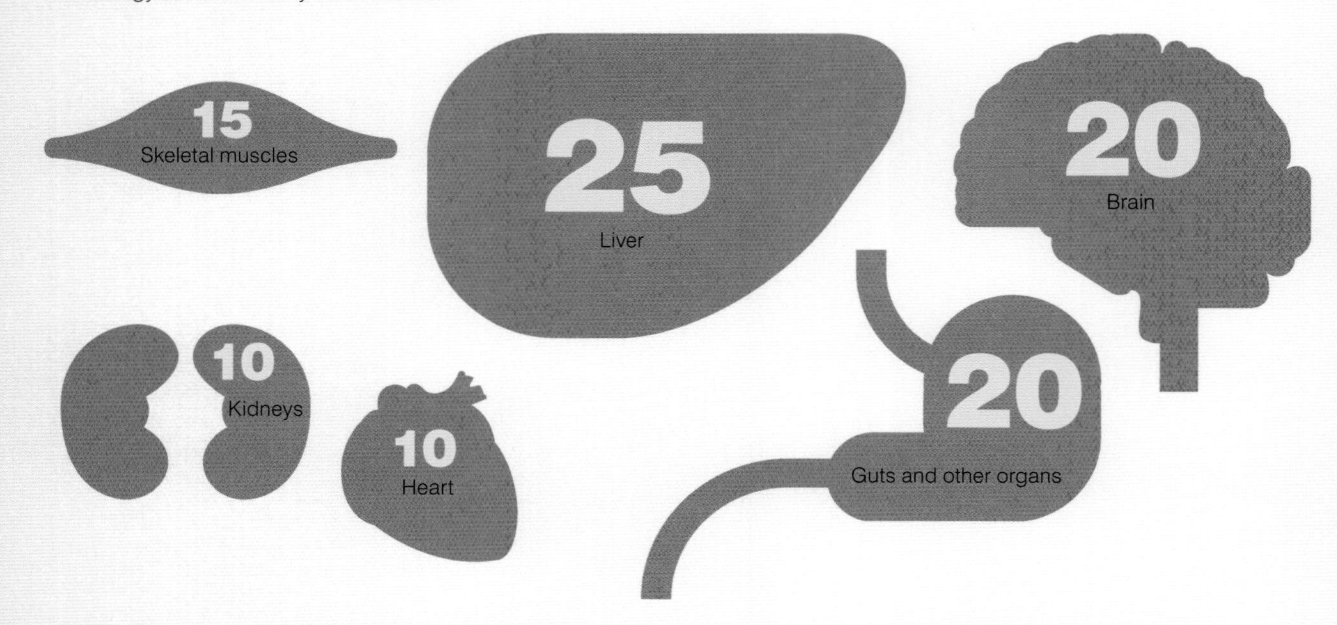

MYSTERIES OF METABOLISM

The term 'metabolism' is a convenient catch-all shorthand for the immense number of chemical reactions, changes and processes occurring throughout every cell of the body, every second of every day, many of which are interlinked and interdependent. Estimates of how many individual chemical reactions happen soon run into millions, then billions, then off the chart. However, the body's use of energy for metabolism has been studied intensively and contributes to many areas of knowledge, from mainstream physiology to sports diets to designing survival rations for extreme conditions.

ENERGY USE (%)

Use of energy (based on factors such as relatively stress-free environment, ambient body temperature, empty stomach).

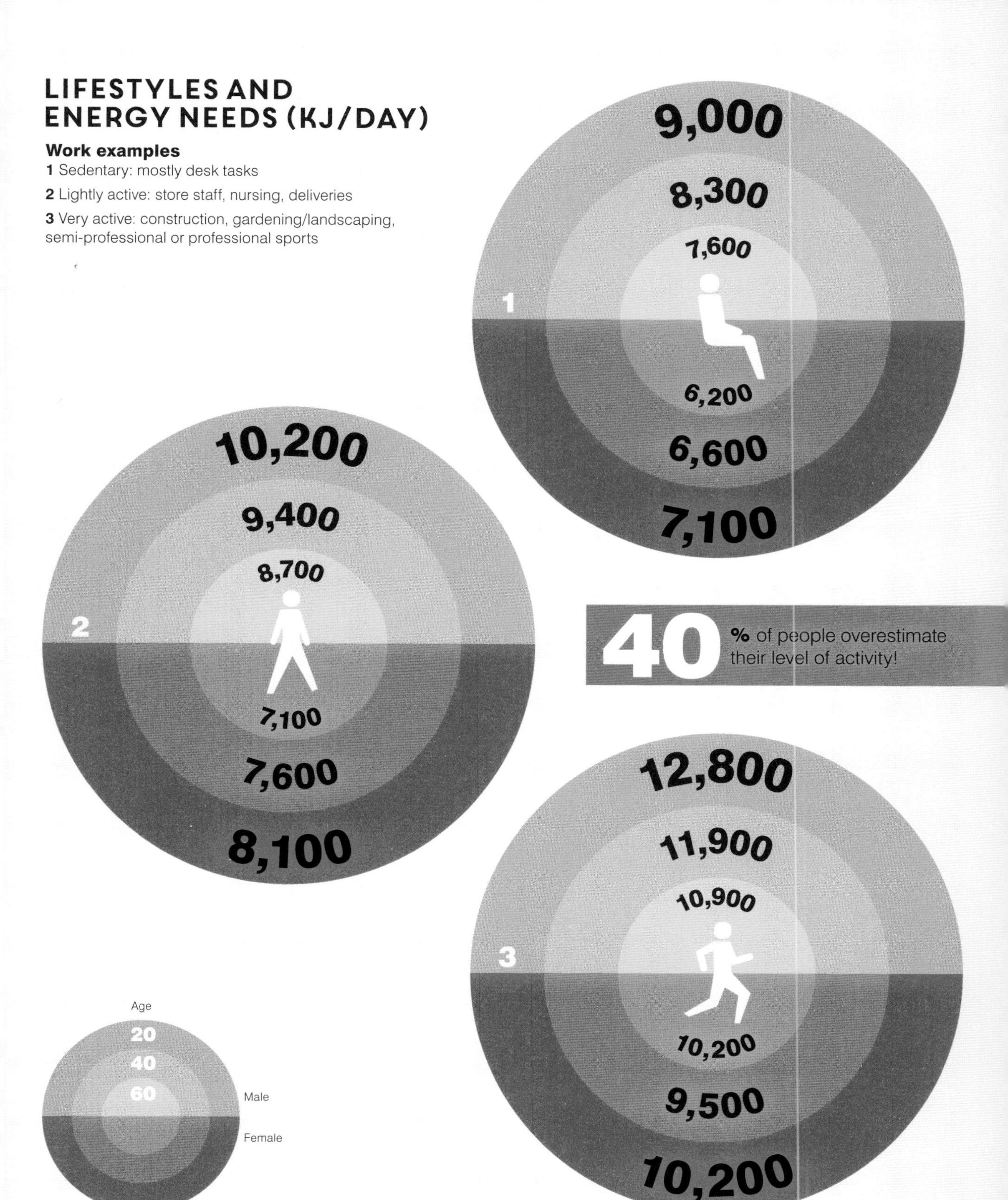

LIFESTYLES AND ENERGY NEEDS (KJ/DAY)
Work examples
1 Sedentary: mostly desk tasks
2 Lightly active: store staff, nursing, deliveries
3 Very active: construction, gardening/landscaping, semi-professional or professional sports
9,000
8,300
7,600
1
6,200
6,600
7,100
10,200
9,400
8,700
2
7,100
7,600
8,100
40 % of people overestimate their level of activity!
12,800
11,900
10,900
3
10,200
9,500
10,200
Age
20
40
60
Male
Female

INS AND OUTS OF ENERGY

The body is an energy transformer. It takes in chemical energy – in the form of trillions of bonds between atoms and molecules in foods and drinks. By the myriad processes of metabolism it converts this energy into other forms, especially the kinetic energy of movement, the thermal energy of heat, the electrical energy of nerve signals, and miscellaneous types such as the sound energy of speech.

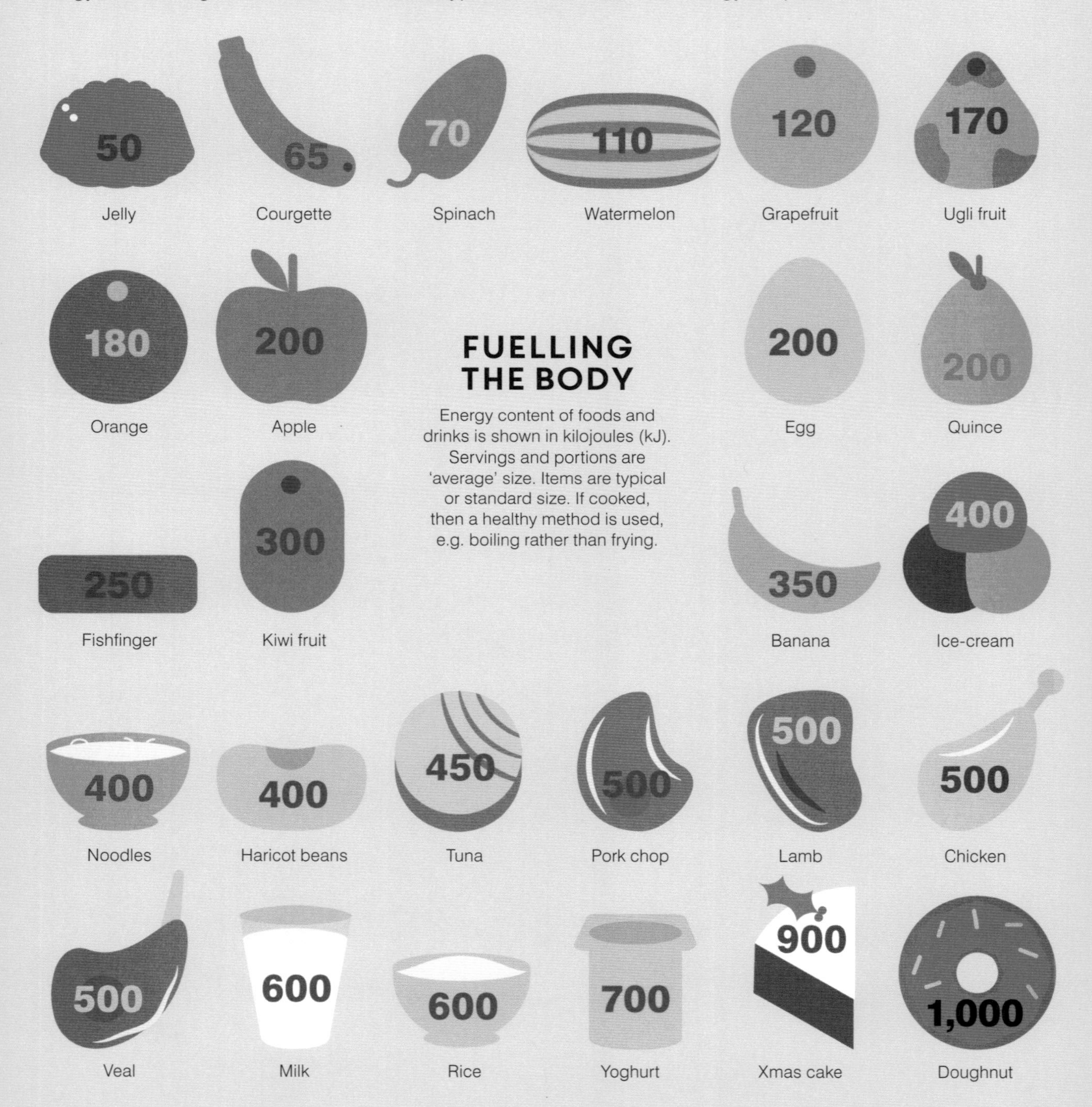

Over time, too much energy coming in and not then used, is eventually converted into body fat. And energy use varies depending on body mass – heavier bodies use more, gender – females use typically 5–10% less than males, and age – expenditure decreases with increasing years. In a typical body, one kilogram of body fat contains enough energy to run between three and four marathons.

DISASSEMBLY LINE

Apart from breathed-in oxygen, every last bit of the human body's energy comes from its food and drink. Acquiring these substances is the task of digestion, and it's an epic story of breakdown and destruction. Every delicious mouthful is chewed to a pulp, slides quickly down the gullet, and is welcomed into a stomach bath of strong acid and destructive juices called enzymes. Now an oozing mush known as chyme, the food is further disassembled by more enzymes in the small intestine, reducing it to molecules miniscule enough to absorb through the intestinal lining into the blood. Next is the large intestine, whose task is to absorb water, some vitamins and other incidentals, before the results come to rest in the rectum ready for removal.

DIGESTIVE AREA

Most absorption of nutrients takes place in the small intestine. It has features that successively increase its inner surface area, compared to a simple tube (shown here in area, sq m).

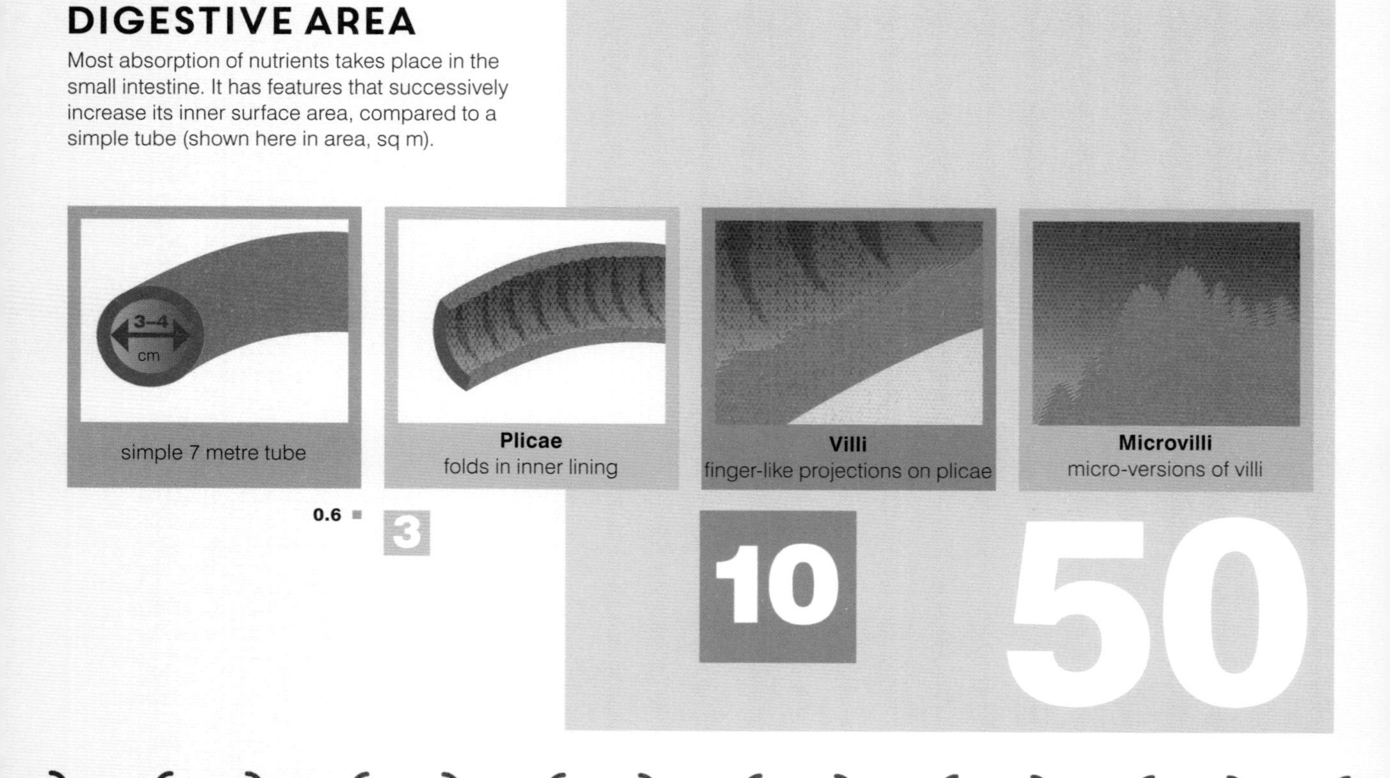

1–1.5
Saliva

1.5–3
Stomach

1–2
Small intestine

1.5–2.5
Pancreas

1
Liver (bile)

0.2–0.5
Large intestine

0.2
Approximately 95% reabsorption means very little water is lost in stools

Digestive juices produced per day (litres)

Digestion involves the production of large quantities of water-based juices, followed by remarkable reabsorption of water in the large intestine. This saves us from having to drink more than 10 litres daily!

1 Assumes healthy, thorough chewing

2 The stomach takes an hour or two longer to tackle fatty foods, compared to carbohydrates and proteins

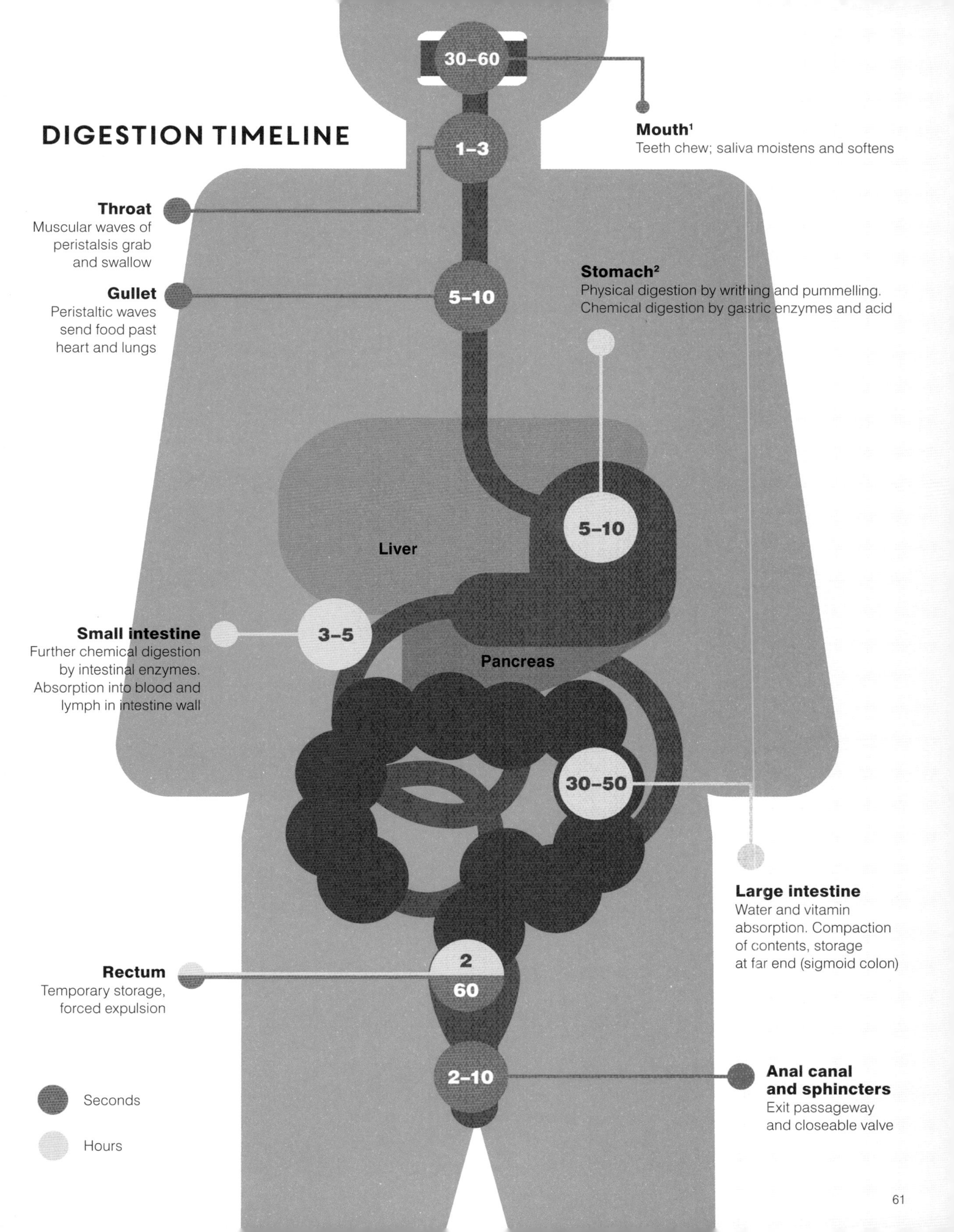
DIGESTION TIMELINE
30–60
Mouth[1]
Teeth chew; saliva moistens and softens
1–3
Throat
Muscular waves of peristalsis grab and swallow
5–10
Gullet
Peristaltic waves send food past heart and lungs
Stomach[2]
Physical digestion by writhing and pummelling. Chemical digestion by gastric enzymes and acid
5–10
Liver
Pancreas
3–5
Small intestine
Further chemical digestion by intestinal enzymes. Absorption into blood and lymph in intestine wall
30–50
Large intestine
Water and vitamin absorption. Compaction of contents, storage at far end (sigmoid colon)
2
60
Rectum
Temporary storage, forced expulsion
2–10
Anal canal and sphincters
Exit passageway and closeable valve
Seconds
Hours

BLOOD CONTENTS

About half of blood is water. The rest is the most vital substances needed for life including dissolved oxygen, energy-rich sugars and fats, antibody proteins to fight disease, and vital nutrients, minerals and vitamins. Delving deep among the red and white cells, numbers and turnover soon become extraordinary. New red blood cells are made at the rate of two to three million every second; each of these cells contains 280 million molecules of the red oxygen-carrier haemoglobin; each haemoglobin has more than 7,000 atoms. That adds up to assembling six million billion atoms per second.

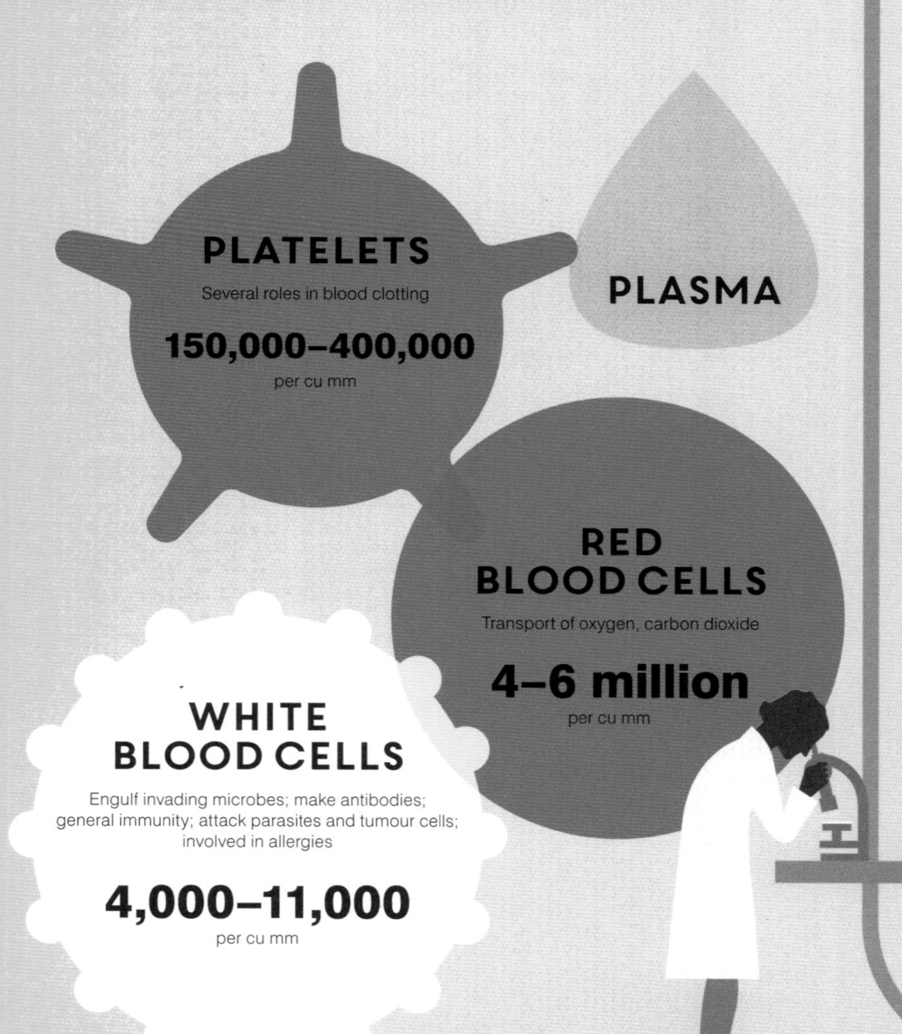

1

0.5

53–57

43–46

MAIN FRACTIONS OF BLOOD

'Fractions' in the sense of relative components or proportions.

Average %

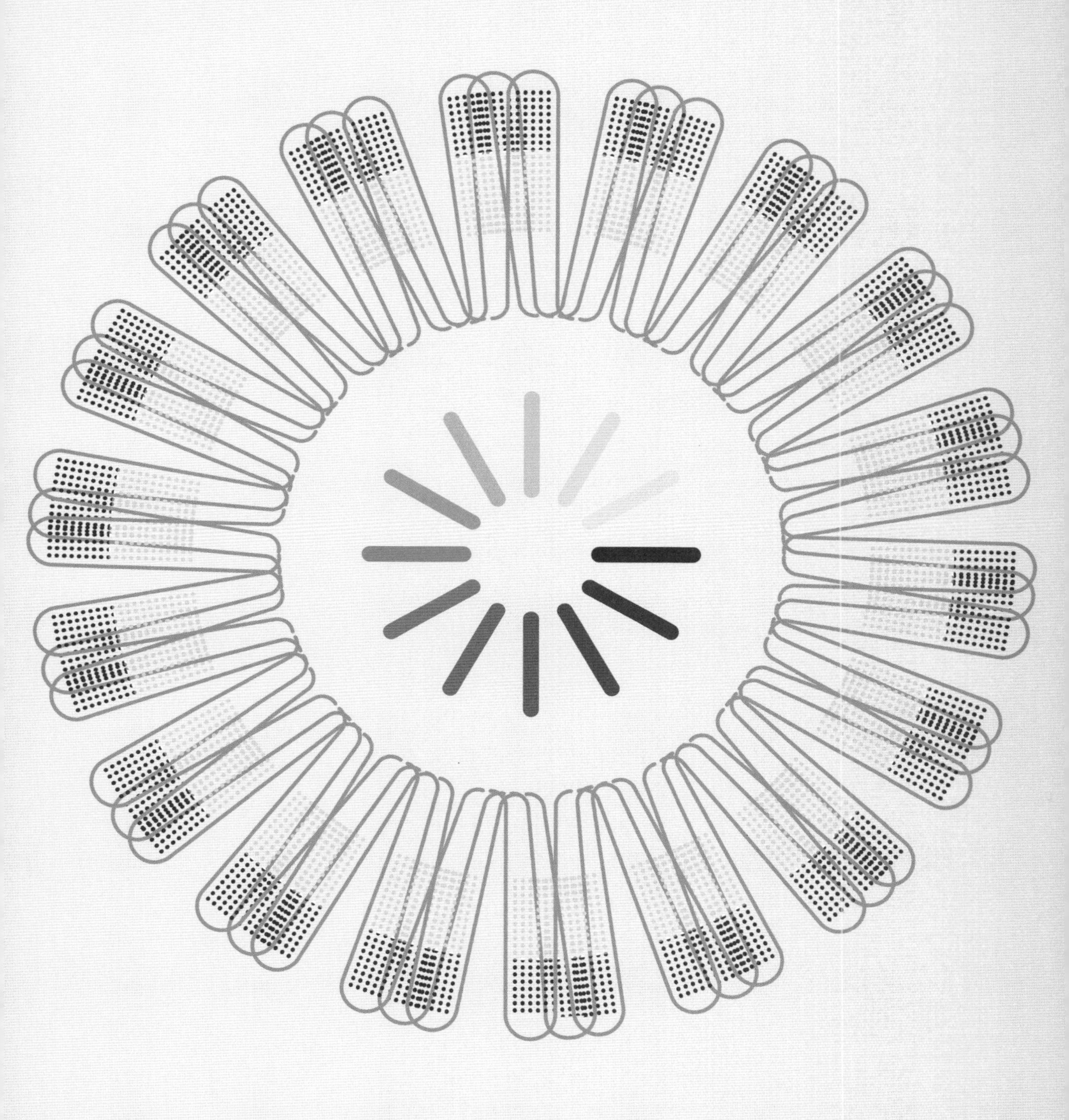

150,000 REVOLUTIONS PER MINUTE

Old-time physicians examined blood by putting it into a test-tube and allowing Earth's natural gravity to separate the components, heaviest at the bottom. Today, fast-spinning ultracentrifuges whirl around blood at more than 150,000 revolutions per minute – 2,500 times per second – producing forces of 2 Mg, two million times normal gravity. This separates blood's tiniest components including viruses, DNA and proteins. Waiting for Earth's gravity to do this would take longer than the predicted age of the Universe.

THE CHEMISTRY OF SURVIVAL

Temperature is a critical factor in the speed of chemical reactions. The body's immense arrays of biochemical activity – its metabolism – are all finely tuned to happen within a very narrow temperature range. This is usually quoted as 36.5 to 37.5°C, with a common variation of up to 1°C over each 24-hour period. Outside this range, the enzymes that control much of the action begin to lose their effects, and soon one disturbed metabolic pathway disrupts another, with rapidly knocking-on results.

DAILY TEMPERATURE VARIATION (°C)

For a typical daytime routine, core body temperature normally rises and falls through each 24 hours in a natural biorhythm. Overlaid on this the core temperature also varies by as much as 0.5°C, depending on surroundings and the body's level of activities.

IN COLD WATER

Depending on water's flow speed, it takes away body heat 25 faster than air. These are approximate times for an averagely competent adult swimmer dressed in ordinary shirt and trousers with a collar-style flotation aid.

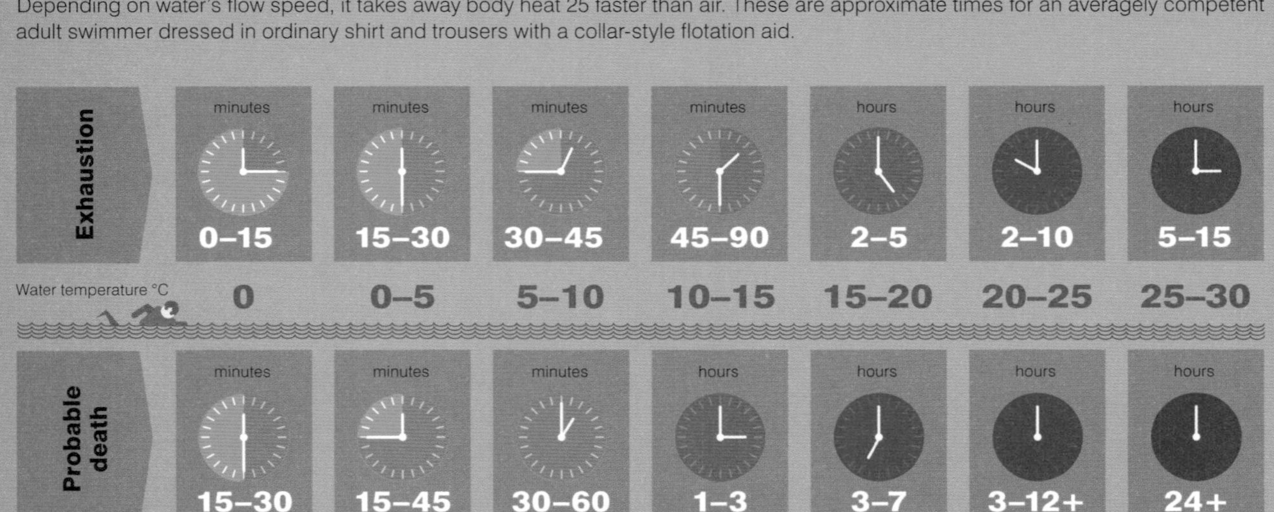

1 Depends on the normal variation of body temperature through the day and night (see above).

PROGRESSION OF HYPOTHERMIA

Deep hypothermia can lead to two strange behaviours:

MILD

32–35 °C

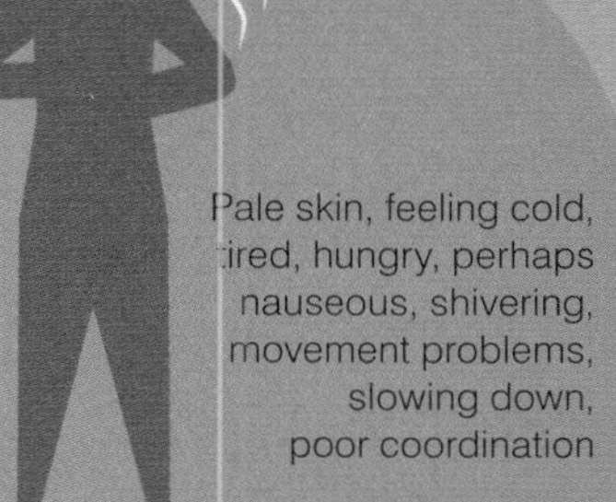

Pale skin, feeling cold, tired, hungry, perhaps nauseous, shivering, movement problems, slowing down, poor coordination

MODERATE

28–32 °C

Slurred speech, feeling disoriented or confused

Breathing and heart rates slow down

SEVERE

Below

28 °C

Motor nerves instruct blood vessels to widen (vasodilation)

Skin and peripheral temperature sensors register excess heat due to warmer blood diverted from core

Brain receives sensations of body becoming too hot

PARADOXICAL UNDRESSING

Removes clothing

Brain registers feelings of vulnerability due to nakedness

TERMINAL BURROWING

Crawls or climbs into enclosed space

Possibly linked to primitive hibernation instinct

LOSS OF CONSCIOUSNESS

GENETIC BODY

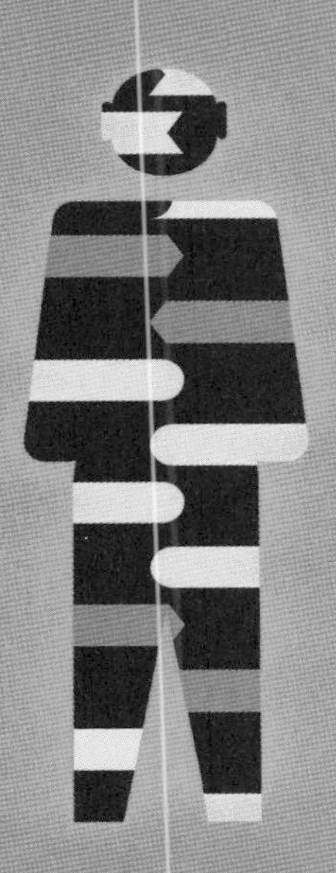

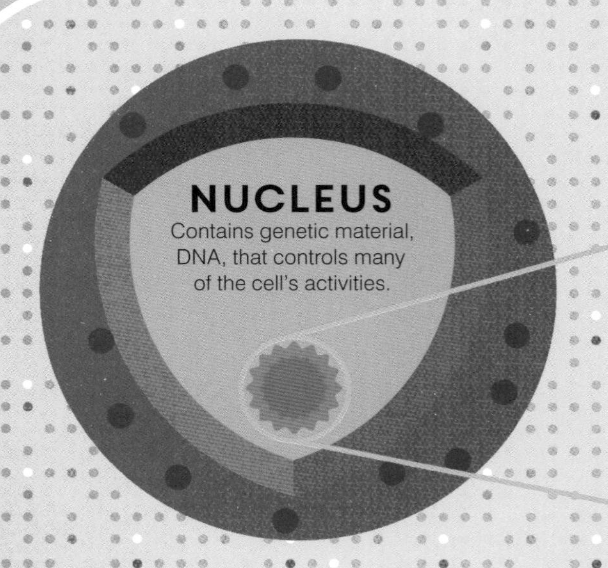

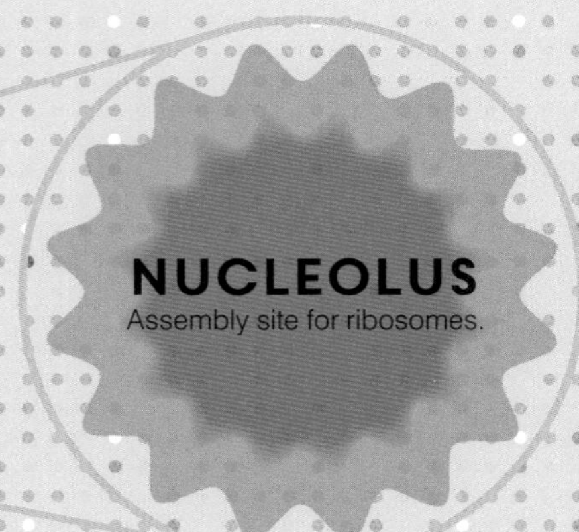

INSIDE A CELL

A 'typical' body cell is shaped like a vague blob about 20 μm across, which means 50 in a row span one millimetre. However there is a snag: the 'typical' body cell does not really exist. Perhaps nearest is the hepatocyte or liver cell, which is a good 'all-rounder' approximately as shown here. Most other cells have very specialized shapes and contents, detailed on the following pages. Just as the body is made of main parts called organs, a cell contains organelles. The largest is usually the nucleus or control centre, which houses the genetic material, DNA. The other main organelles are also shown here, with their main functions.

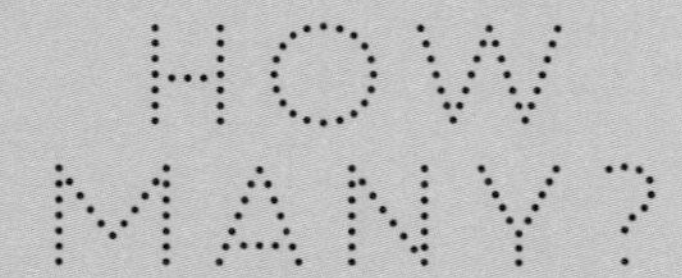

Estimates for the number of cells in the body have **ranged from a few billion to 200,000 trillion (200,000,000,000,000,000)**.

Estimates using cell volume give **15 trillion**, and weight, **70 trillion**.

A recent calculation takes into account cell sizes, numbers and how they are packed in different tissues.This estimate gives **37 trillion cells (37,000,000,000,000)**.

Counting them at one per second would take slightly more than **one million years**.

CELL MEMBRANE
Controls what enters and leaves the cell and protects the interior.

HOW HEAVY?

The typical weight for a cell is 1 nanogram

Which is one billionth of a gram or...

0.000,000,001

grams

CYTOPLASM
Provides the cytoskeleton for the cell's shape, internal framework and organization; contains dissolved substances.

MITOCHONDRION
Breaks apart high-energy substances, for example sugars to provide the cell's energy.

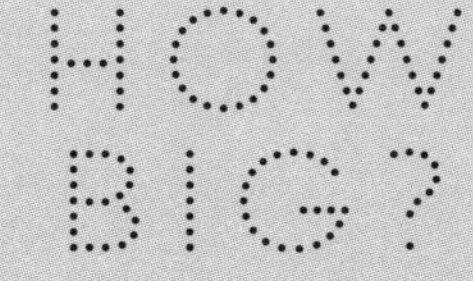

LYSOSOME
Breakdown and recycling area for old, unwanted substances.

ENDOPLASMIC RETICULUM
Synthesis of lipids, protein processing, enzyme storage, detoxification.

HOW BIG?

The average size or volume for a human or other mammal cell is

0.000,004

cubic mm

Which is 4 billionths of a cubic cm.

RIBOSOME
Protein synthesis – joins amino-acid subunits to make larger molecules and proteins (see page 76).

CELLS GALORE

There are more than 200 different kinds of cells in the body. Each is specially shaped, and has its own complement of parts and organelles inside, to fulfil its particular role. For example, a nerve cell or neuron has long, snaking projections – the axon (fibre) and dendrites – to communicate with its fellows. Muscle cells are packed with mitochondria since they require plentiful energy, while red blood cells are lilttle more than bags of oxygen-transporting haemoglobin. The examples below list some of their unique characteristics.

SKIN

Keratinocytes
Flattened, filled with keratin for hardness and protection.

RED BLOOD

Erythrocytes
'Biconcave' shape with large surface area to absorb oxygen.

WHITE BLOOD

Leucocytes
Flexible to enable squeezing among tissues in pursuit of invaders.

SKELETAL MUSCLE

Striated myocytes
Long and spindle-shaped, able to shorten when contracting.

HEART MUSCLE

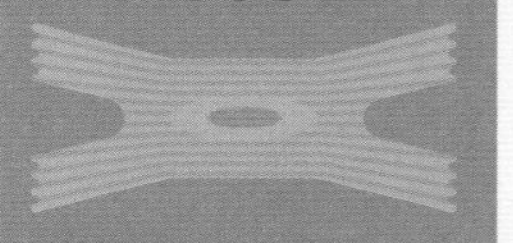

Cardiomyocytes
Branching and interlaced; some work while others rest.

NERVE

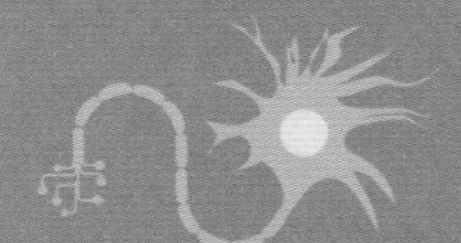

Neurons
Many thin extensions to interconnect with other nerve cells.

FAT

Adipocytes
Large bag-like vacuole of stored fat.

BONE

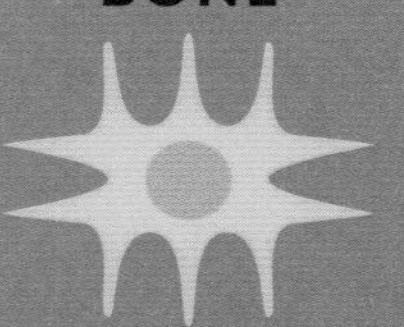

Osteoytes
Spider-shaped to maintain and repair bone around them.

INSULIN-MAKING

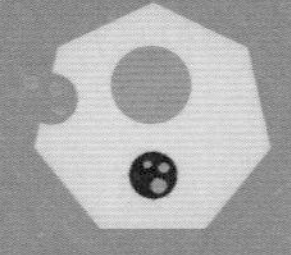

Pancreatic ß cells
Contain many receptacles of insulin hormone.

GOBLET

Columnar epitheliocytes
Produce mucus in the gut, airways and other sites.

SCHWANN

Neurolemmocytes
Manufacture mylein to surround and protect nerve fibres.

CONNECTIVE TISSUE

Fibroblasts
Many branches to manufacture collagen and other connective substances.

It's estimated that bacteria and other microbes living in and on the body, most of them 'friendly', outnumber body cells by **10 to 1**, that is, totalling around **400** trillion which is **2,000** times more than stars in our galaxy, the Milky Way.

2
Heart

60
Skin

50
Fat deposits

BILLIONS OF CELLS...

240
Liver

2,000
Brain

DOWN AMONG THE DNA

Within the nucleus, or control centre, of a human cell reside 46 lengths of the genetic substance DNA or to give it its full name, deoxyribonucleic acid. Each DNA length, plus its associated protein substances called histones, is known as a chromosome. These chromosomes occur in 23 pairs, with the two members of the pair being approximate copies of each other. These lengths of DNA carry, in the form of a chemical code, the genes – instructions for how the body and all its parts develop, work, maintain and repair themselves.

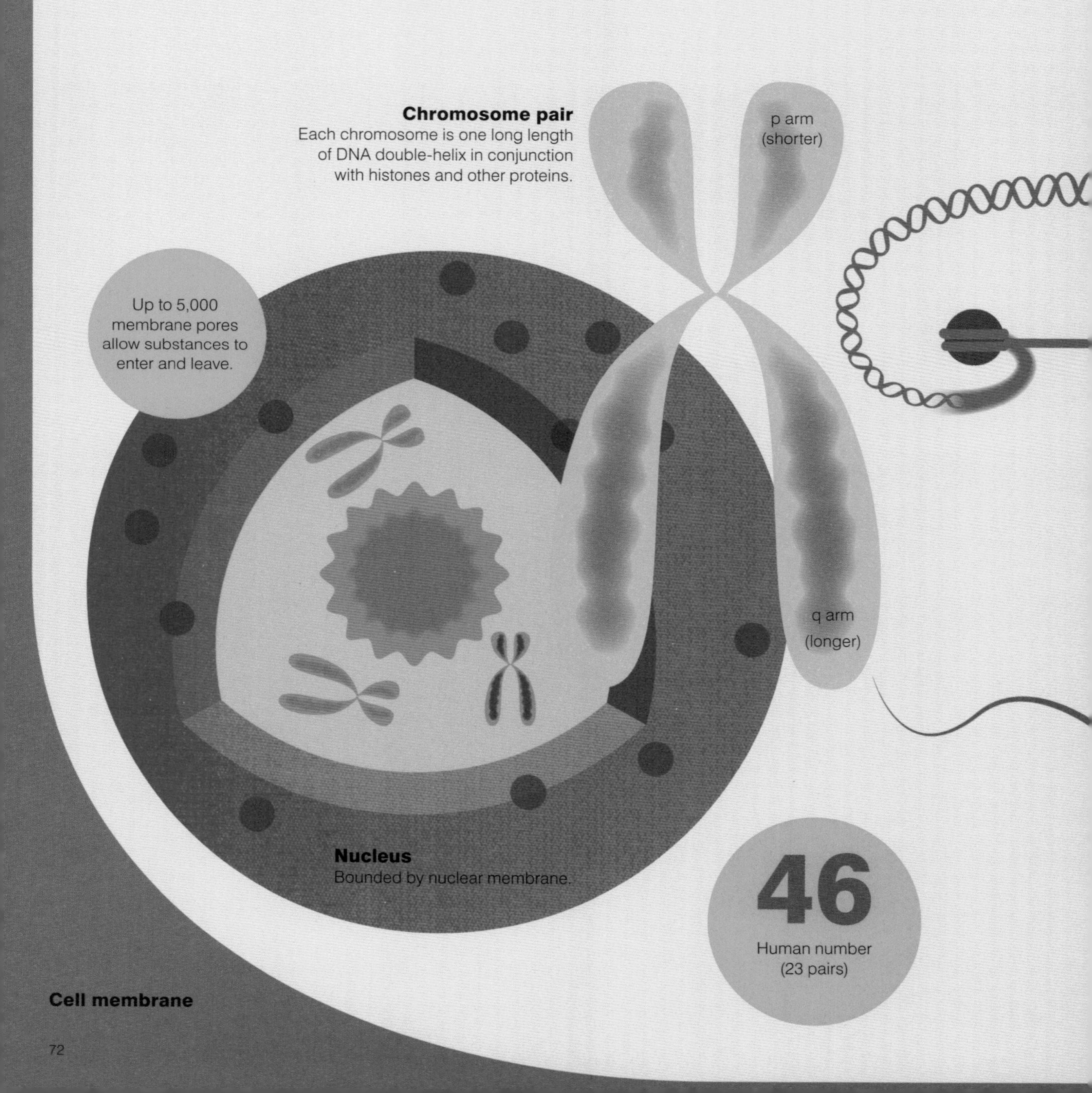

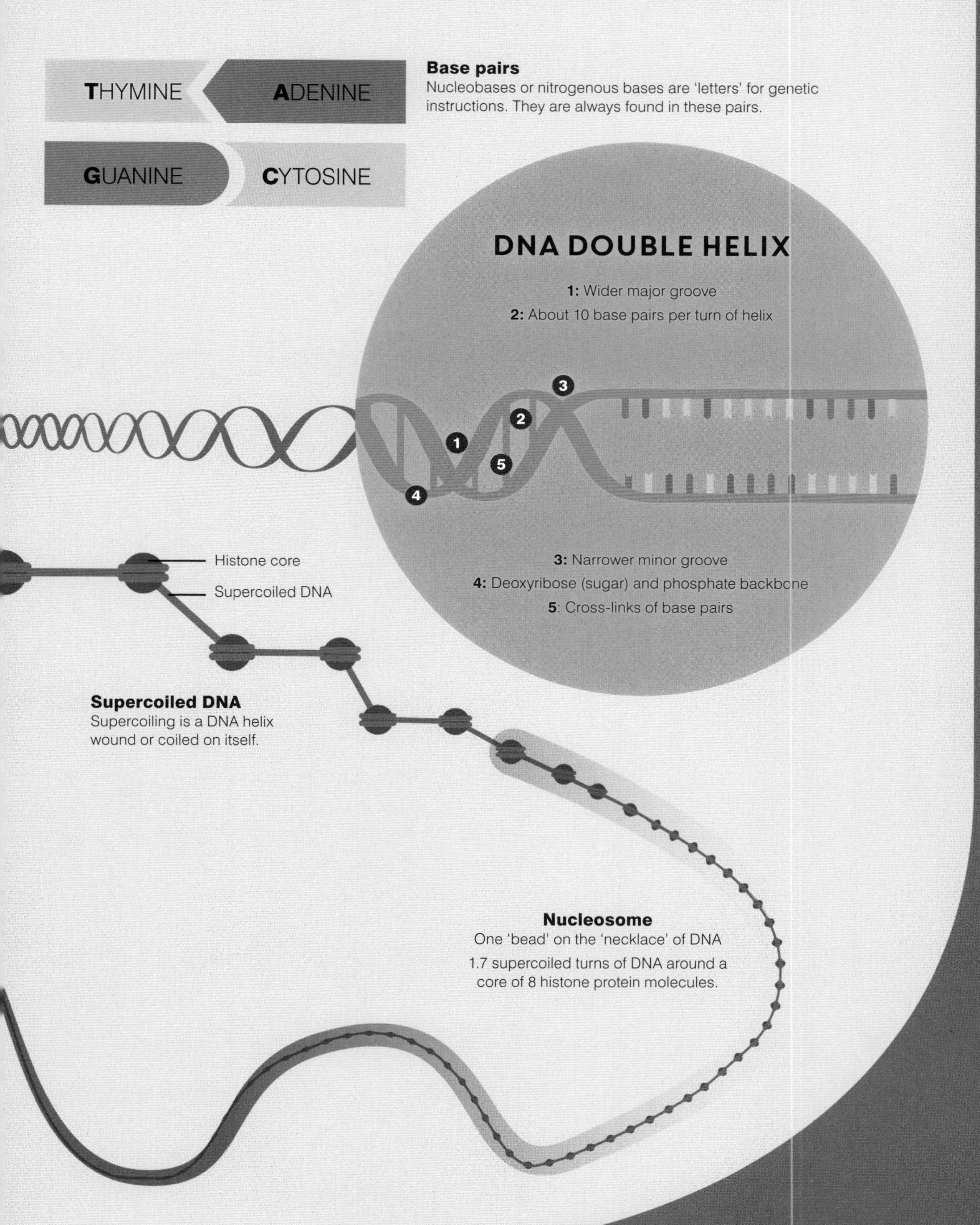
THYMINE
ADENINE
GUANINE
CYTOSINE
Base pairs
Nucleobases or nitrogenous bases are 'letters' for genetic instructions. They are always found in these pairs.
DNA DOUBLE HELIX
1: Wider major groove
2: About 10 base pairs per turn of helix
3: Narrower minor groove
4: Deoxyribose (sugar) and phosphate backbone
5: Cross-links of base pairs
1
2
3
4
5
Histone core
Supercoiled DNA
Supercoiled DNA
Supercoiling is a DNA helix wound or coiled on itself.
Nucleosome
One 'bead' on the 'necklace' of DNA
1.7 supercoiled turns of DNA around a core of 8 histone protein molecules.

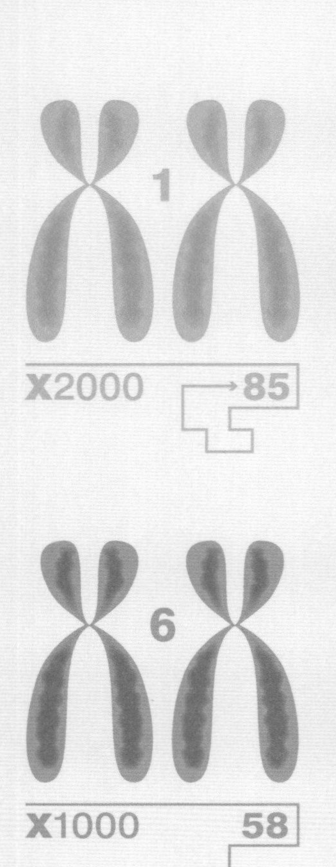
1
X2000 85
6
X1000 58

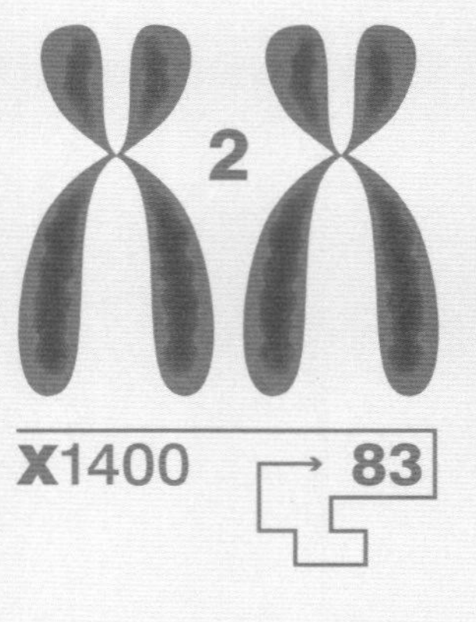
2
X1400 83

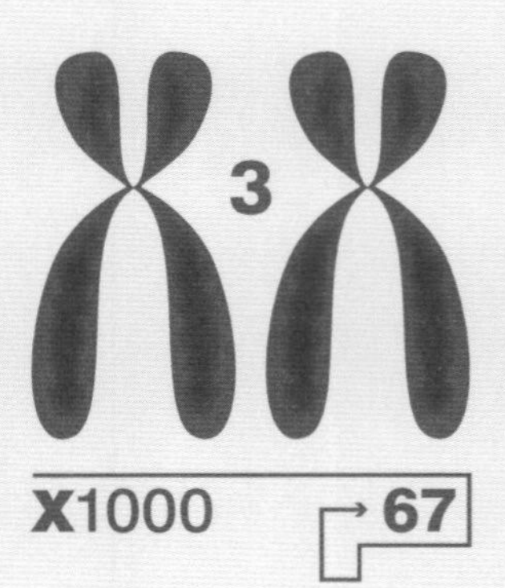
3
X1000 67

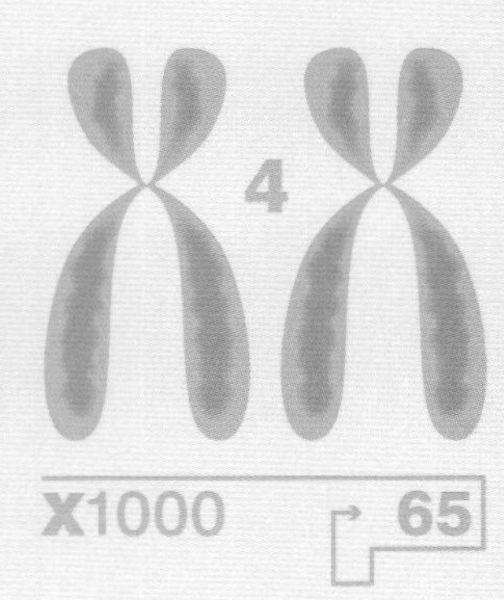
4
X1000 65

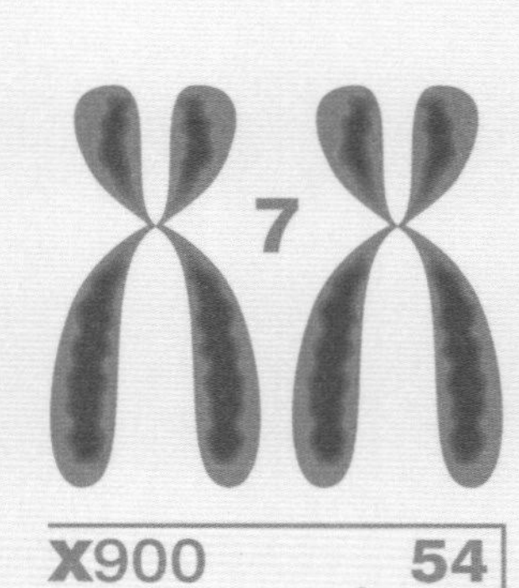
7
X900 54

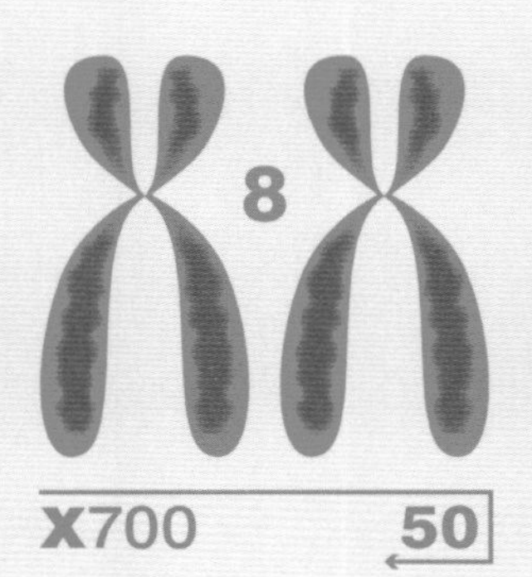
8
X700 50

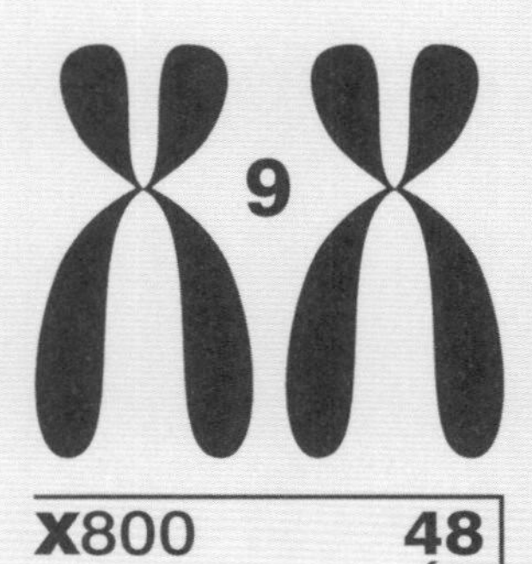
9
X800 48

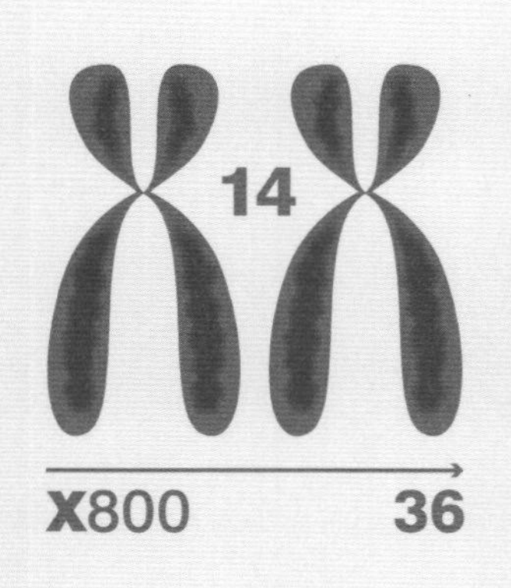
14
X800 36

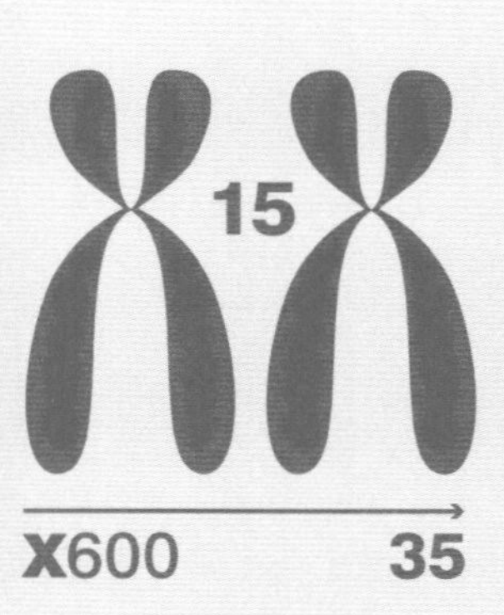
15
X600 35

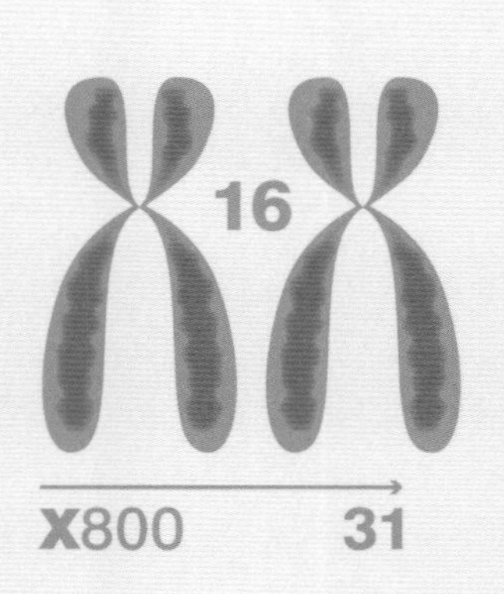
16
X800 31

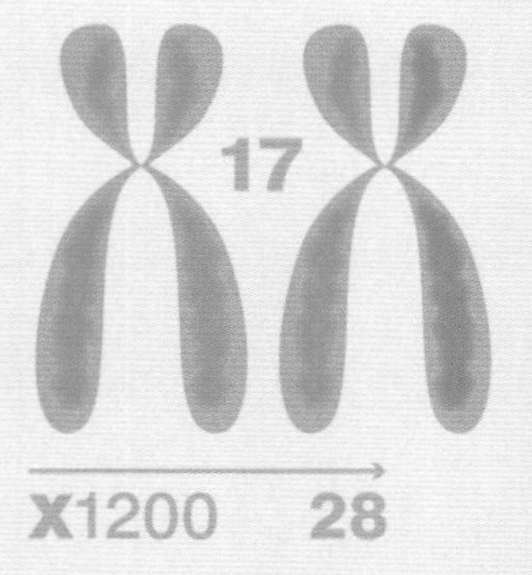
17
X1200 28

22
X500 17

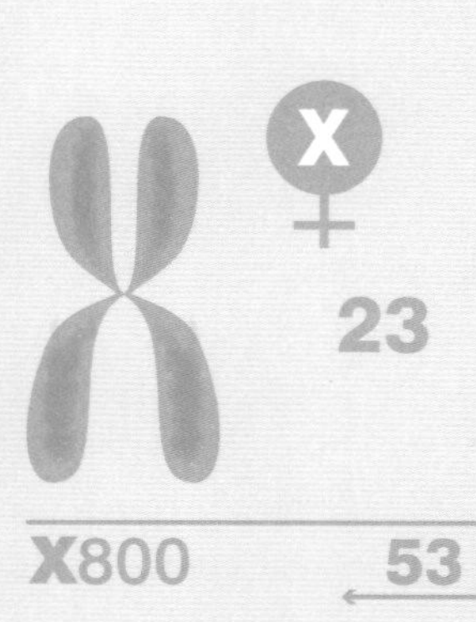
X
23
X800 53

Y
X50 20

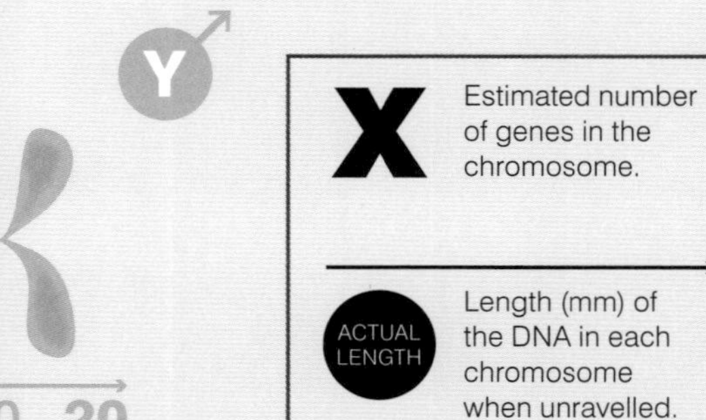
X
Estimated number of genes in the chromosome.
ACTUAL LENGTH
Length (mm) of the DNA in each chromosome when unravelled.

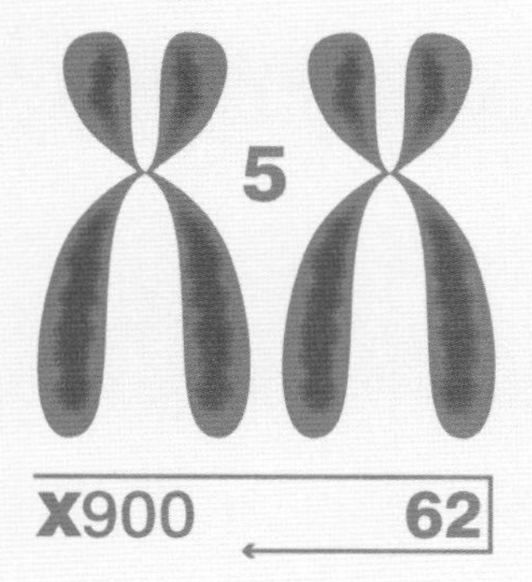

KARYOTYPE

A karyotype is the appearance of all the chromosomes in a particular organism, usually laid out in a row. In the human karyotype there are...

22 pairs of identical-looking chromosomes, numbered approximately in order of decreasing size.

The 23rd pair are dissimilar and known as **X** and **Y** sex chromosomes.

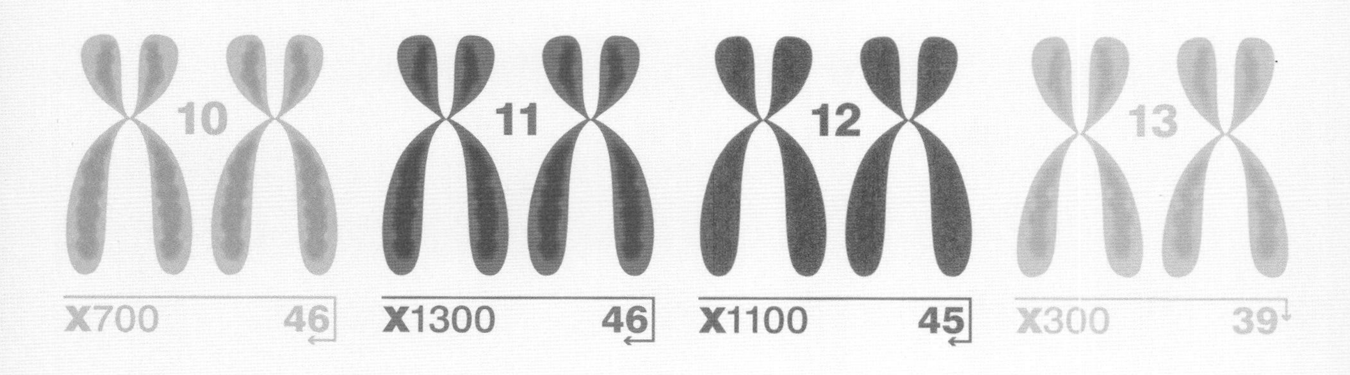

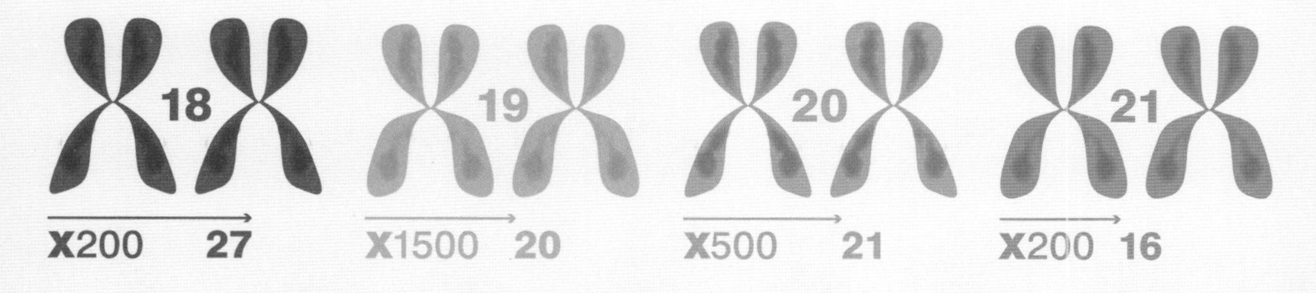

THE GENOME

The full set of genetic instructions for the body is known as the human genome. It is carried in the cell nucleus by 46 strands of DNA double-helix, each one immensely long in molecular terms, but far too thin to see under a light microscope. However, as the cell prepares to divide, each snaking DNA length twines and twists onto itself into supercoils and supersupercoils. Eventually it takes up a shorter, thicker, condensed, X-like form that, with a suitable stain (dye), is visible under the microscope. These items became known as chromosomes, 'coloured bodies' – a term that applies whether they are in condensed X-shapes ready for cell division (in duplicated pairs shown here), or strung out and meandering as they dispense instructions.

HOW GENES WORK

Genes are directions for how the body develops and works. But what do they actually do? Like a blueprint plan or instruction book, a gene is a length of DNA that holds the information, as a chemical code, to build a part of the body. This part is usually molecular in scale. For many genes, the parts are proteins, such as actin and myosin, which power muscles; collagen and keratin that toughen skin; amylase and lipase and other digestive enzymes, and hundreds of others. Further genes direct the construction of different kinds of RNAs, ribonucleic acids, which are heavily involved in organizing and running the cell's activities – including control of its genes.

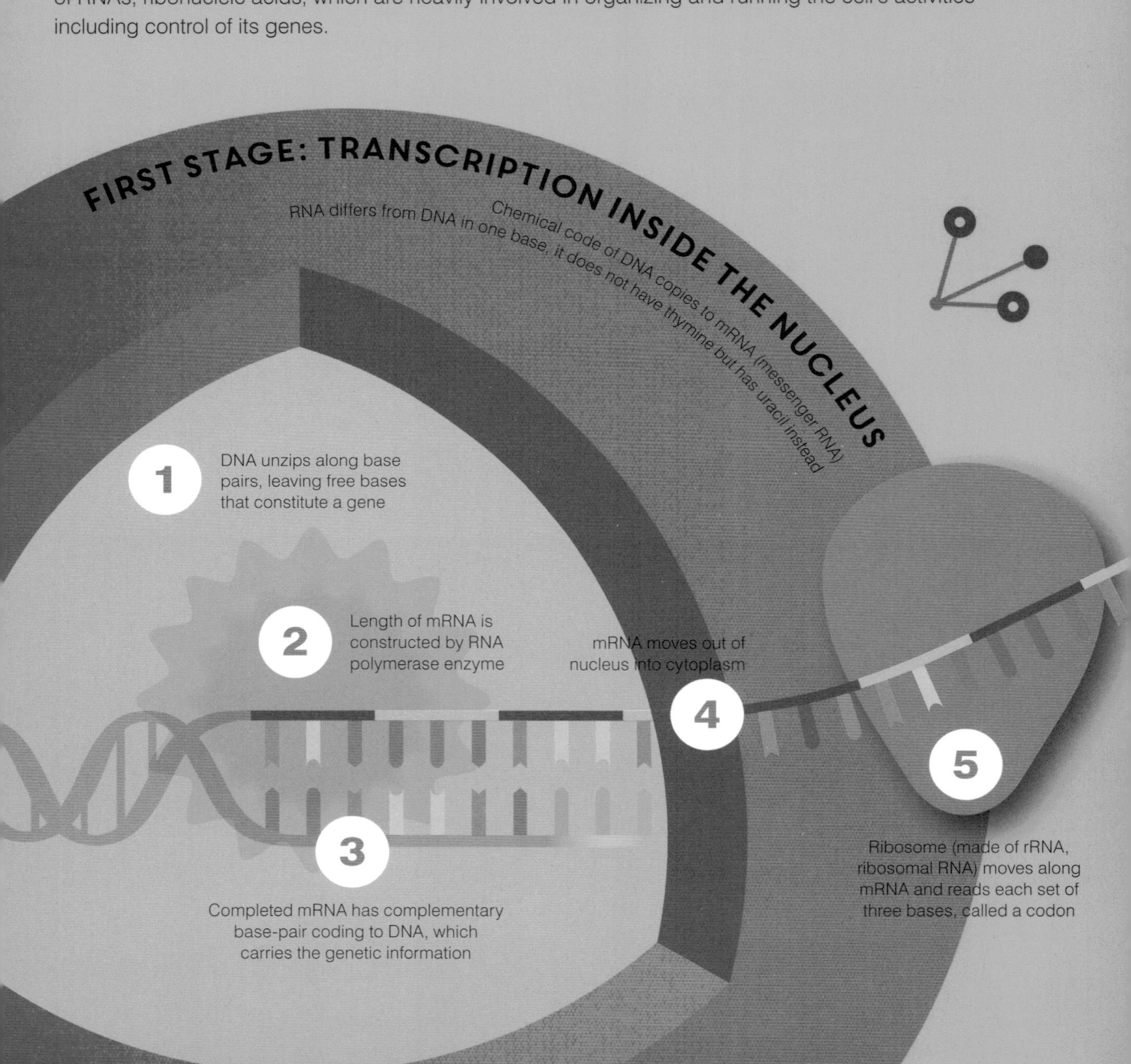

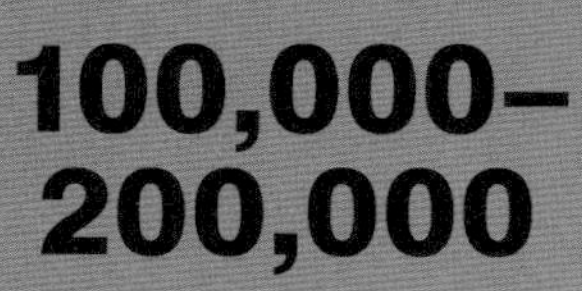

100,000–200,000

Different proteins in the body

20,000

Estimated number of genes carrying information to make proteins

20

Different amino acids in all living things. Linking them in different sequences produces the variety of proteins

SECOND STAGE: TRANSLATION IN THE CYTOPLASM

mRNA's coded information is used for protein synthesis, assisted by a ribosome and tRNAs (transfer RNAs)

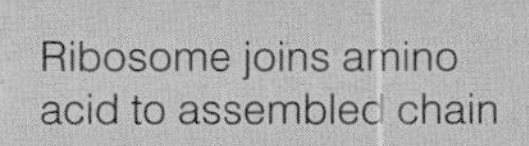

7 Ribosome joins amino acid to assembled chain

6 tRNA brings correct amino acid specified by the codon

8 Chain of amino acids lengthens to form a protein

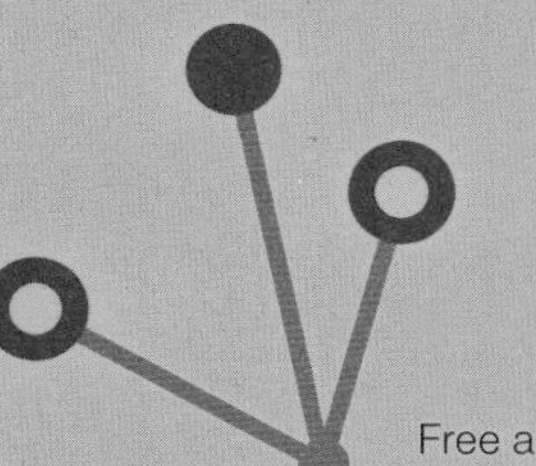

Free amino acids

HOW GENES SPECIALIZE

Each cell has a full and complete set of genes – so how do different cells take on their varied appearances and functions? Answer: not all the genes are active or switched on. Usually, essential 'housekeeping' genes are working for basic functions like building organelles and dealing with energy and wastes. But most other genes are switched off or suppressed – apart from those for the cell's specific function. For instance, a red blood cell has its 'housekeeping' genes working, and also those to make oxygen-carrying haemoglobin, while most of the rest are suppressed.

STAGE 1: THE GENETIC INFORMATION

Chromosome 11
Haemoglobin subunit beta gene, HBB.
Location 11p15.5 (chromosome 11, short or p arm, position 15.5).

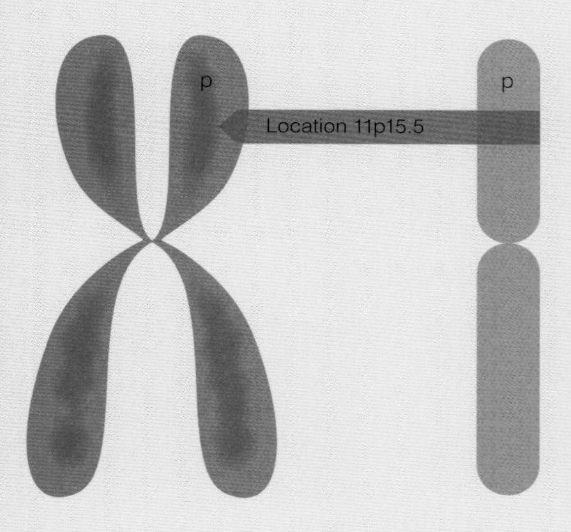

Chromosome 16
Haemoglobin subunit alpha 1 gene, HBA1.
Haemoglobin subunit alpha 2 gene, HBA2.
Location 16p13.3 (chromosome 16, short or p arm, position 13.3).

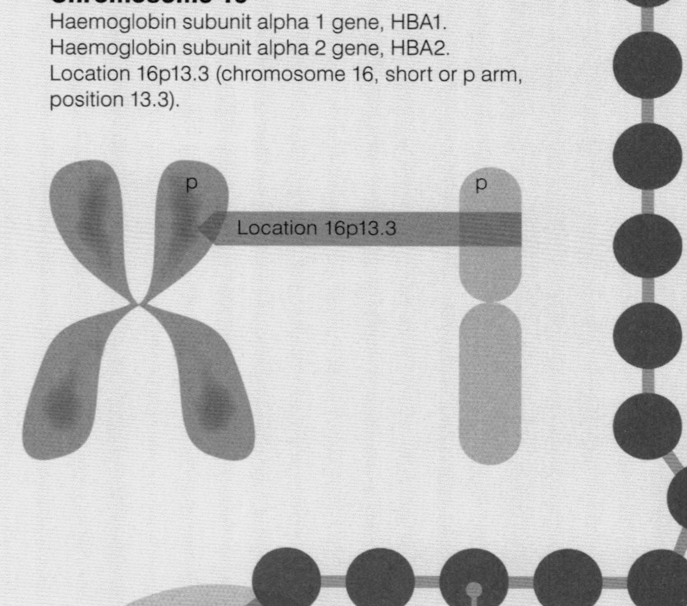

STAGE 2: MAKING THE PROTEIN SUBUNITS

Production of beta globin chain.

mRNA assembled using HBB as template

mRNA is 'read' by ribosome, which assembles amino acid building-blocks.

Chromosome 11

DNA unzipped to expose HBB gene.

STAGE 3: ASSEMBLING THE HAEMOGLOBIN MOLECULE

Primary structure
Sequence of 146 amino acids for one beta globin chain.

Secondary structure
Sequence takes on bends and kinks due to angles of bonds between amino acids, forming an alpha helix.

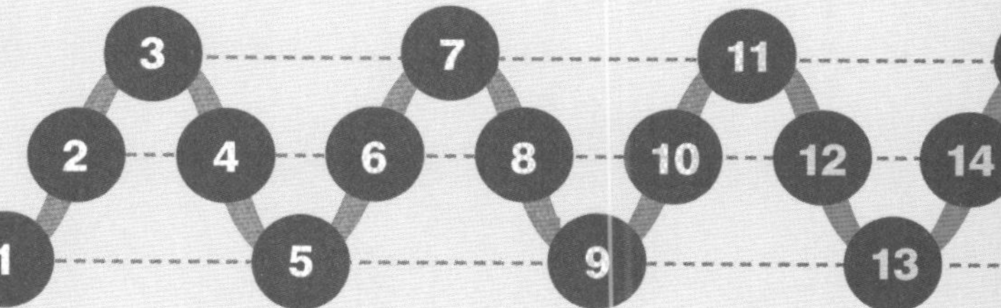

Tertiary structure
Long chain of amino acids (polypeptide) takes on folds, loops and sheets to give full three-dimensional shape of beta globin.

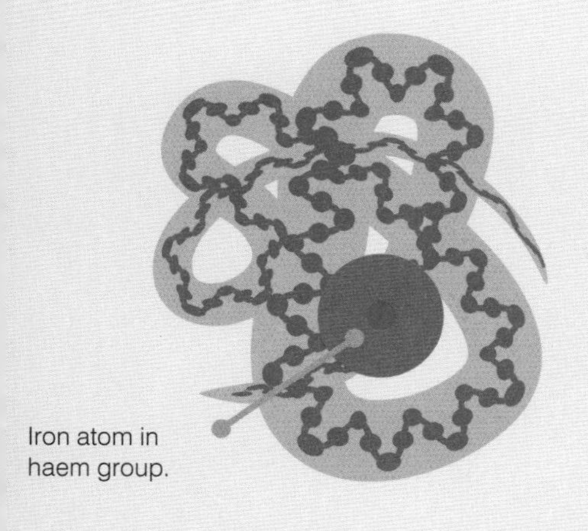

Iron atom in haem group.

Quaternary structure
Alpha, beta and other subunits are assembled to produce complete, fully-working haemoglobin protein.

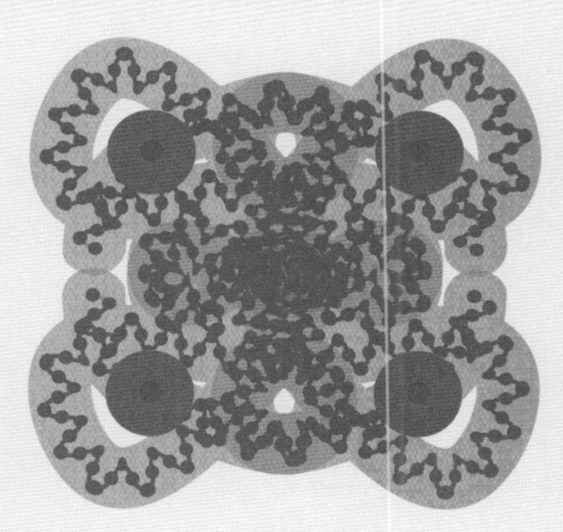

HAEMOGLOBIN WITHIN A RED BLOOD CELL:

280,000,000

Haemoglobin molecules per red blood cell

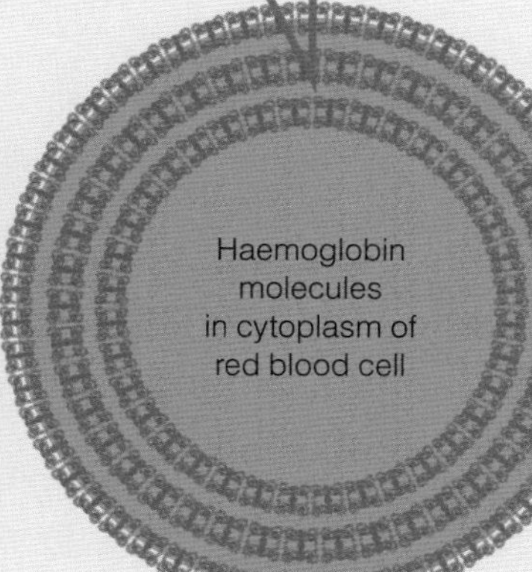

1/3

of the volume of a red blood cell is haemoglobin

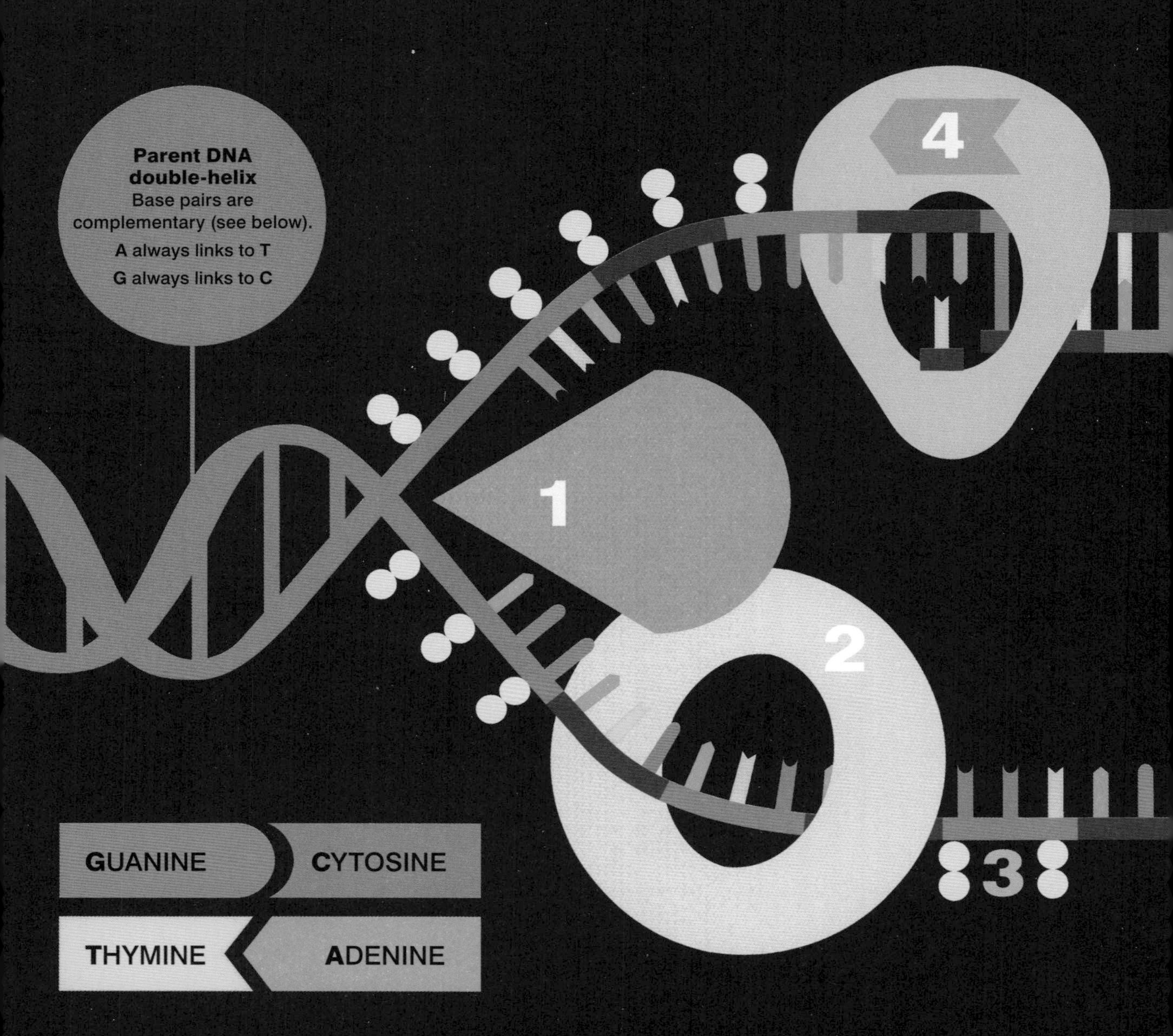

DNA DOUBLE-UP

No cells live for ever. They divide to produce offspring cells, as shown on the next page. The key to this is copying or replicating the genes, which are chromosomes consisting of lengths of DNA, deoxyribonucleic acid. This allows each offspring cell to receive its full set of genes and continue its parent's work. DNA replication underlies almost every process and event in the body, starting from the amazing development of the body from its first set of DNA in its first single cell, the fertilized egg, through to everyday cell division to replace skin, blood and other worn-out cells.

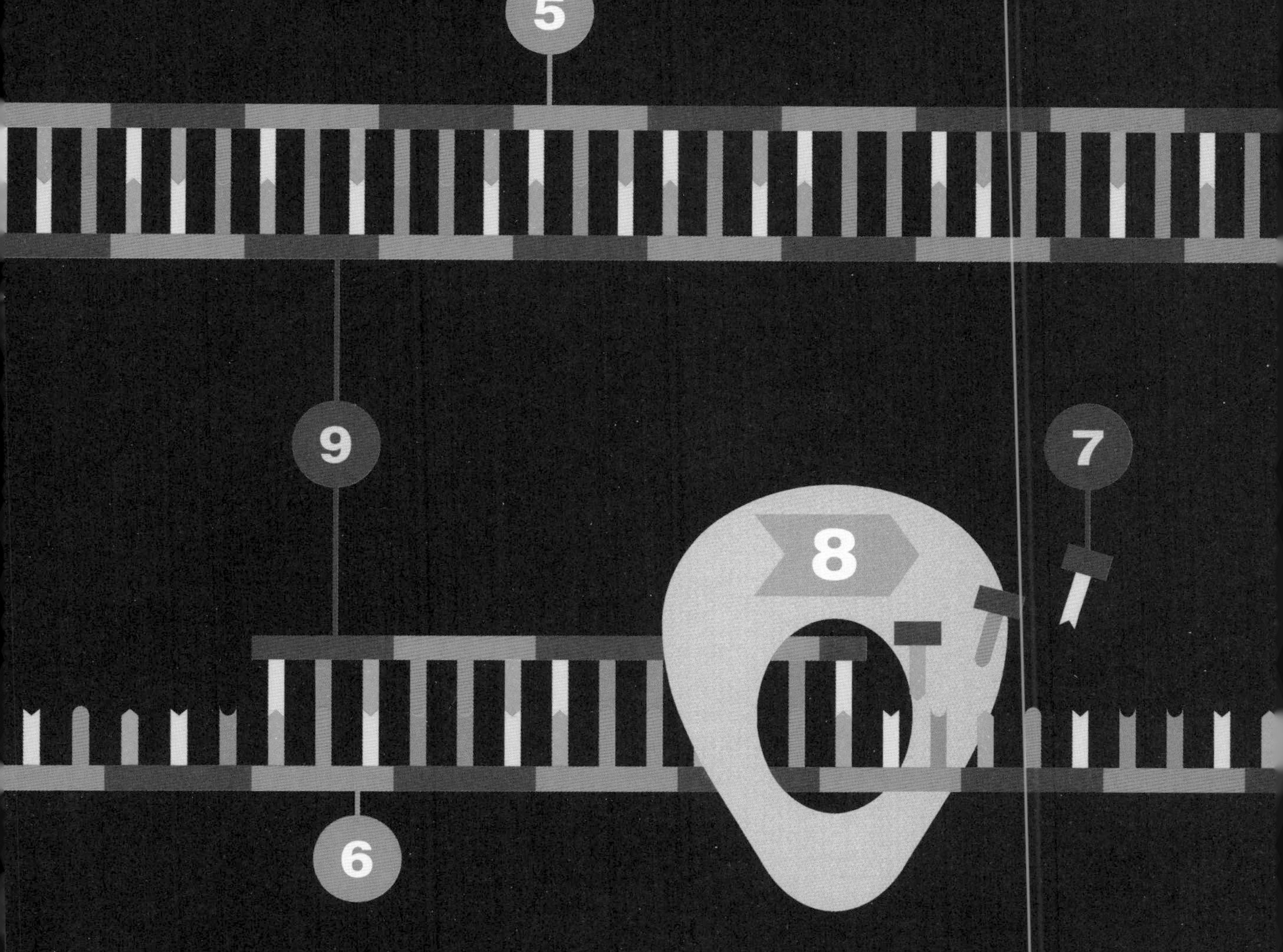

1: Helicase
Enzyme to unwind and separate or 'unzip' the two existing or parent DNA strands at the links between the bases.

2: Primase and RNA primer
Primase enzyme makes RNA primer, the start point for making the new complementary partner or offspring DNA strand.

3: Binding proteins
Protect the exposed bases to stop them re-linking, detaching or degenerating.

4: DNA polymerase
Enzyme that 'reads' existing bases, and 'clips' new bases, sugars and phosphates together to form new complementary strand.

5: Leading strand
The existing DNA strand along which DNA polymerase moves continuously to form the lengthening new strand.

6: Lagging strand
DNA polymerase only works in one direction along the DNA backbone so goes 'backwards' in a step-like manner on this strand.

7: Okazaki fragments
Short sections of DNA newly made on the existing DNA lagging strand which will be joined by DNA ligase.

8: DNA polymerase and DNA ligase
'Backstitch' Okazaki fragments together to form one long new complementary partner on the existing lagging strand.

9: Offspring DNA
Two identical double-helix lengths, each with one parent DNA strand and one new complementary DNA partner strand.

HOW CELLS DIVIDE

Cells do not arise spontaneously from non-living matter. (Except, as biologists theorize, more than three billion years ago when they first evolved.) Instead, each cell comes from a pre-existing cell by the process of cell division – sometimes known, confusingly, as cell multiplication. This nearly always produces two cells from one, as the original or parent cell gives rise to two offspring or sibling cells. The key to such division is splitting of the nucleus, known as mitosis. Prior to this all of the genetic material, DNA, is duplicated so that each offspring cell receives a full set. (Division to make the sex cells, eggs and sperm, is slightly different – see page 180.)

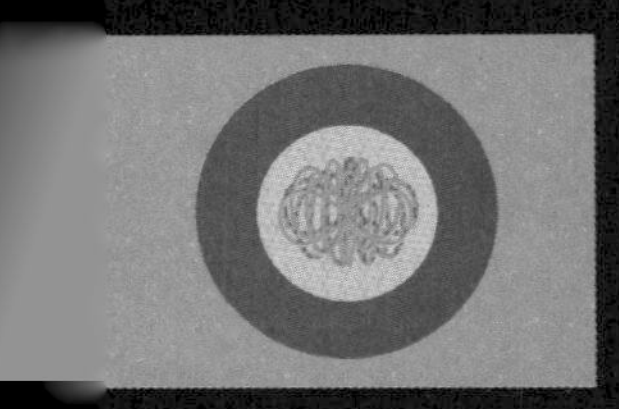

INTERPHASE
DNA of chromosomes spread out and winding, genes active. Also DNA duplicates.

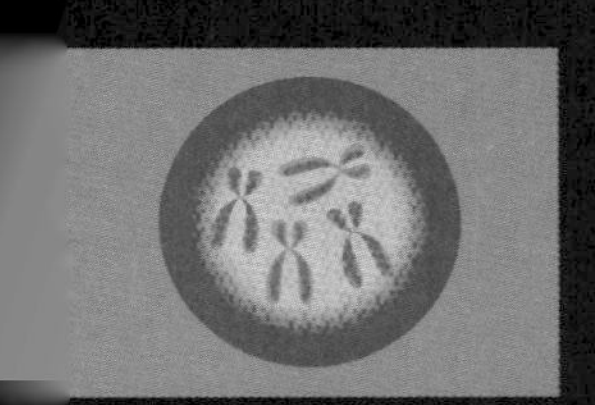

PROPHASE
DNA of each chromosome coils and 'condenses' to become visible. Nuclear membrane disintegrates. Spindle forms from centrosomes and microtubules.

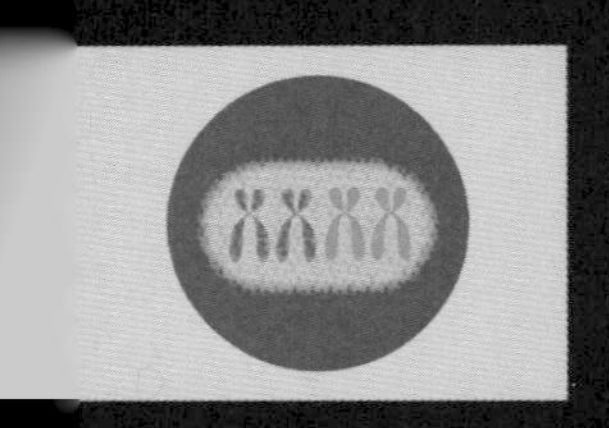

METAPHASE
Microtubules attach to chromosomes. Chromosomes align at centre or equator of cell.

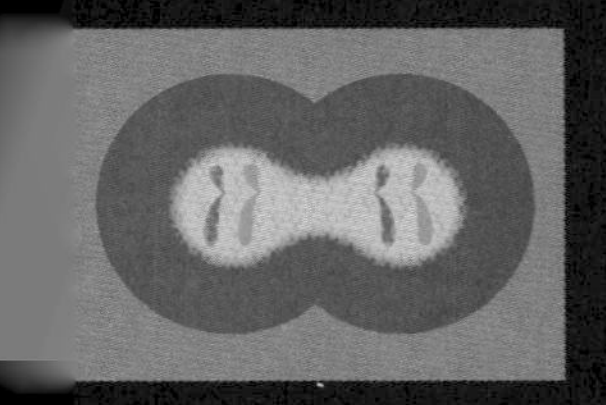

ANAPHASE
Pairs of duplicated chromosomes separate, pulled by microtubules to opposite ends of the cell.

TELOPHASE
Chromosomes reach their positions in each sibling cell. Nuclear membrane reforms in each sibling cell.

80

10

4

1

3–7

CELL DIVISION

Figures shown average % of a cell's life spent in each stage.

CYTOKINESIS

Division of the whole parent cell into two sibling cells. Timing varies but may start early in mitosis. Contractile ring around the middle pinches the cell forming a cleavage furrow. Two sibling cells finally become independent units.

THE LIVES OF CELLS

Each of the body's 200-plus types of cells has its own programed time for existence, before being replaced with more of its kind from that tissue's rapidly multiplying stem cells. In general, hard physical wear or chemical exposure means faster turnover. Deep in the brain are the longest survivors – the neurons that give us thoughts, feelings and memories. As a rough guide to their prodigious numbers, the cells turned over in one second in the body, laid end to end, would stretch more than one kilometre.

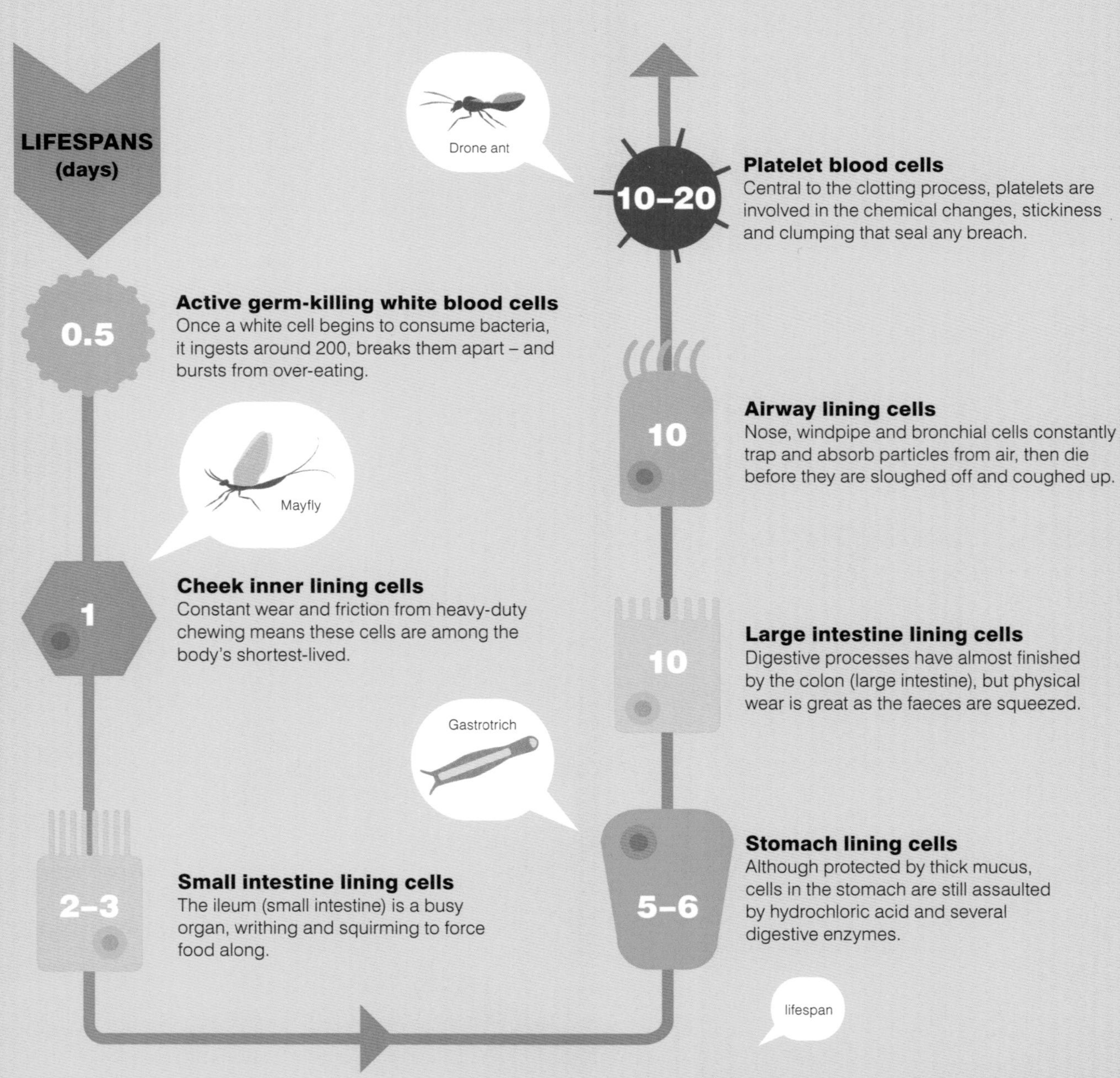

10–20

Retinal cells, eye
Average lives of the light-sensing or photoreceptor cells, rods and cones, mean a constant slow turnover in the delicate eye lining.

20–30

Epidermal (outer skin) cells
Physical wear, friction and minor injuries mean the whole outer layer of skin, the epidermis, replaces itself at least monthly.

120

Red blood cells
Bone marrow produces more than two million every second, as the same number's minerals are recycled especially by the spleen and liver.

150

Liver cells
Known as hepatocytes, liver cells are multi-taskers, handling all kinds of minerals and nutrients, as well as storing vitamins.

350
(1 year)

Pancreatic cells
Some pancreatic cells make the hormones insulin and glycogen, others produce digestive enzymes for the small intestine.

30,000
(80 years)

Brain neurons
Their enormously complex architecture, with thousands of synapses (connections), mean cerebral neurons can last almost a lifetime.

22,000
(60 years)

Memory white cells
After an infection, a few memory T and B cells circulate for years, even decades, ready to spring into action and fight the same disease again.

10,000
(25 years)

Bone maintenance cells
Osteocytes have complex shapes, like a three-dimensional spider with over 100 'legs'. They keep bone minerals topped up and turning over.

5,500 (15 years)

Skeletal muscle cells
Muscle cells or myocytes are big 'multicells' made of many smaller cells fused into one unit with diameters up to one millimetre.

500
(16 months)

Lung lining cells
The tiny air sacs, alveoli, accumulate bits of dust and other debris at a slow rate and so are replaced every year or two.

HOW GENES INTERACT

The human genome has 46 chromosomes, or lengths of DNA, as 23 pairs. That is, there are two chromosome 1s, two 2s, and so on. Does this mean two identical copies of each gene, one on each chromosome of the pair? Like most of genetics, the answers are yes, no and maybe. In some people the two versions, or alleles, of a particular gene are identical. In others the two alleles are different. One is stronger or dominant, and it 'overpowers' its weaker or recessive counterpart. An example is the gene for the blood group Rhesus, RH. It has alleles for being Rhesus positive, RH+, or Rhesus negative, RH–. Oh, and much much more (see right).

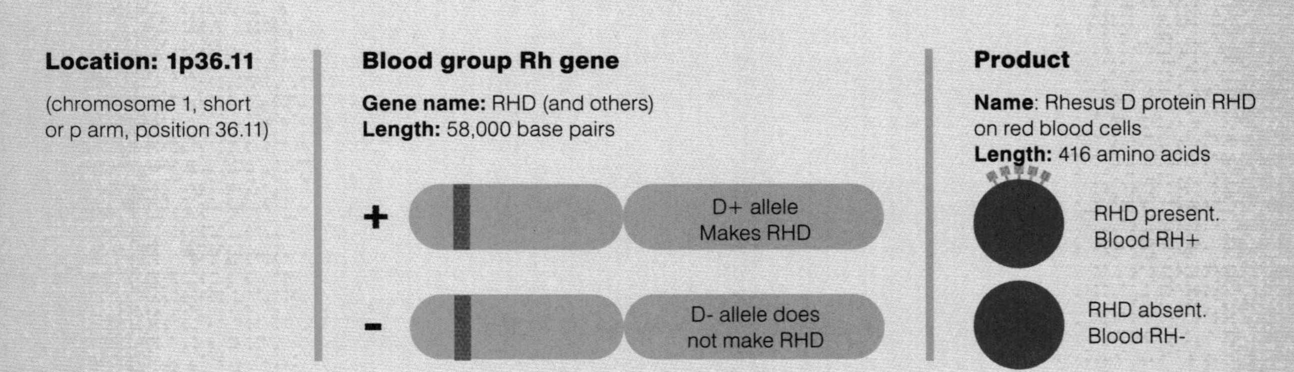

The three possibilities
The three possible combinations of RHD genes depend on the alleles on the two chromosome 1s. One of these comes from the mother and the other from the father. D+ is stronger or more dominant, D– is weaker or recessive.

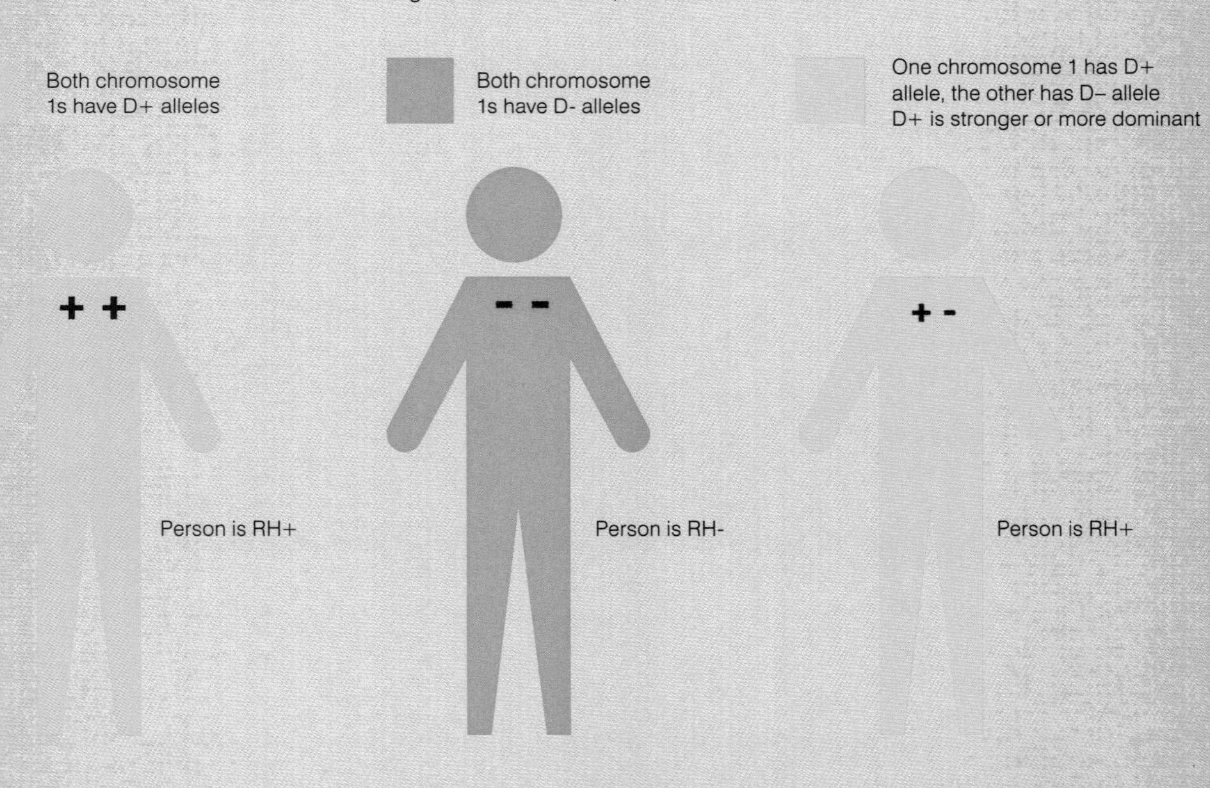

GENETICS IS NEVER THAT SIMPLE

The explanation here of the **Rhesus blood group** is very simplified.

There are not just two alleles for the RHD gene, but more than 50.

This means there are many RHD proteins, like Weak D, Partial D, DEL and several more.

And not all Weak Ds are the same. There's Weak D type 1, Weak D type 2, Weak D type 4, Weak D type 11, Weak D type 57 and more.

On top of all this, RHD is only one of several genes in the 'rhesus family'.

Others are RHCE, RHAG, RHBG and RHCG, some on different chromosomes.

These make a variety of other proteins such as C, E, c and e.

And remember that Rhesus is only one blood group. There are also ABO on chromosome 9, MNS on chromosome 4, L (Lewis), K (Kell) – more than 30 groups in all.

That's just one tiny glimpse of why genetics is so complicated.

INHERITING GENES

Genes are inherited directly from our parents. As mentioned, each cell in the body has two full sets of genes, as pairs of chromosomes 1 to 23. These are replicates of replicates of replicates many times over, via cell division, from the original first two sets of chromosomes. One of these original sets was in the mother's egg cell and the other in the father's sperm cell (see page 182). See how different combinations of different versions, or alleles, of a gene produce different results – but smile first!

DIMPLES

These small cheek pits or indentations are probably caused by a dominant version or allele of the dimples gene, let's call it **+**. Not having dimples is the recessive allele, **–**. Remember that only one of the mother's two dimples genes can go into each egg, likewise one from the father into each sperm. How they combine is all down to chance.

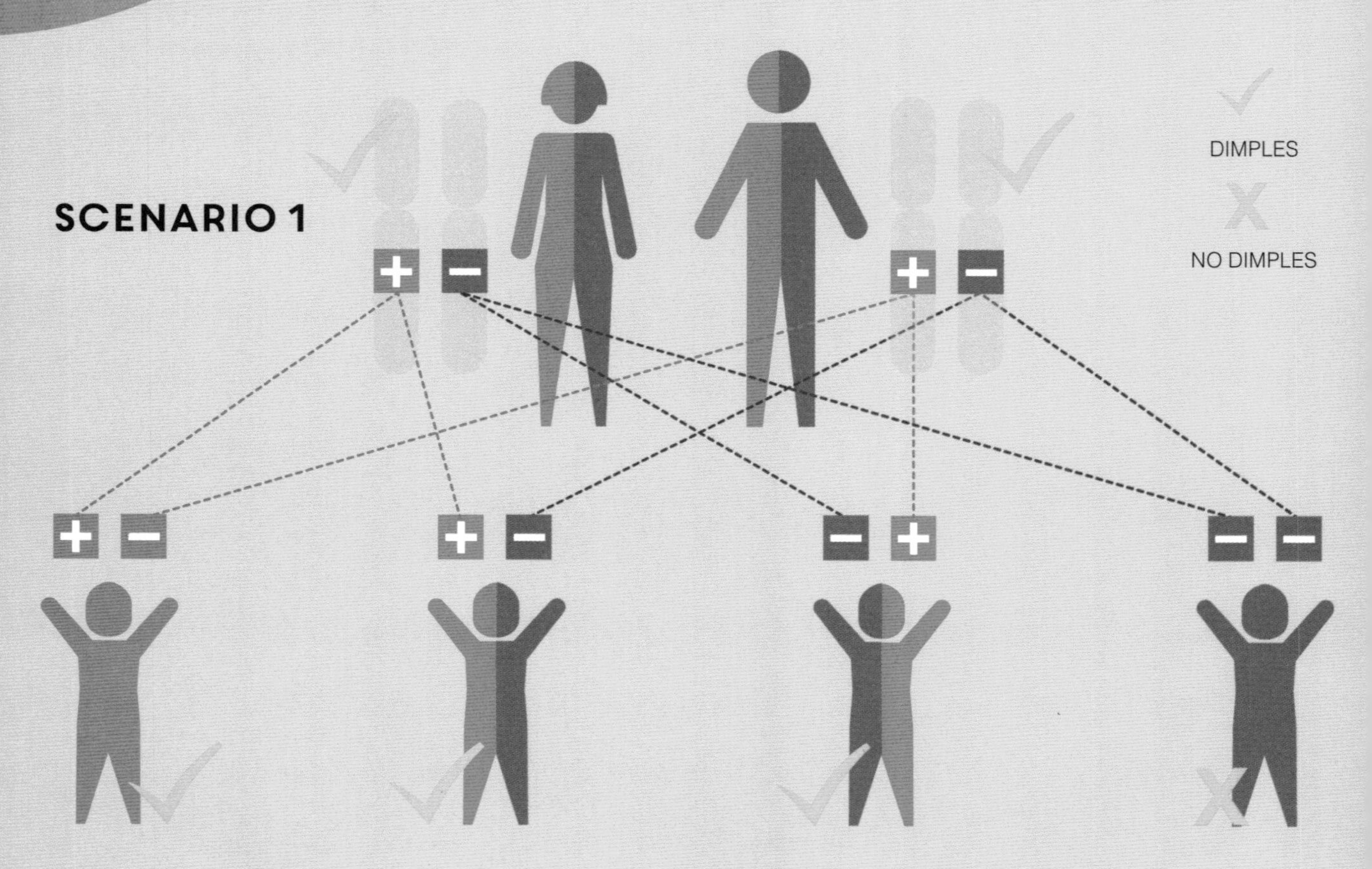

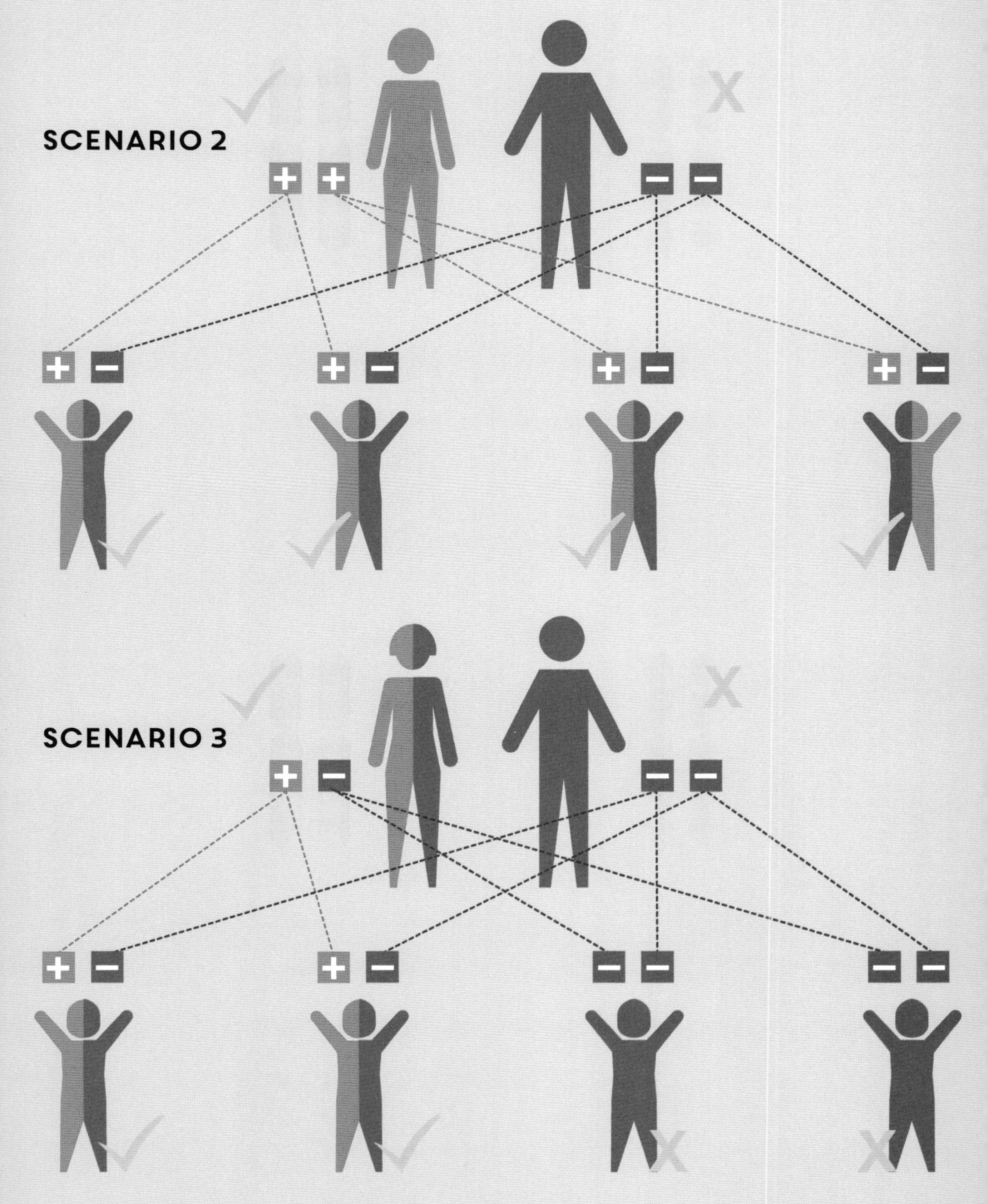
SCENARIO 2
SCENARIO 3

THE GENETIC EVE

In a cell, each 'powerpack' mitochondrion has short lengths of DNA, called mDNA or mtDNA. When a sperm joins an egg at fertilization, it does not contribute mitochondria. So all of a body's mtDNA came solely from the mother. Studies of mtDNA changes or mutations trace our human species, ***Homo sapiens***, back to a theoretical female 'Genetic Eve' (Mitochondrial Eve) in Africa 200,000 years ago.

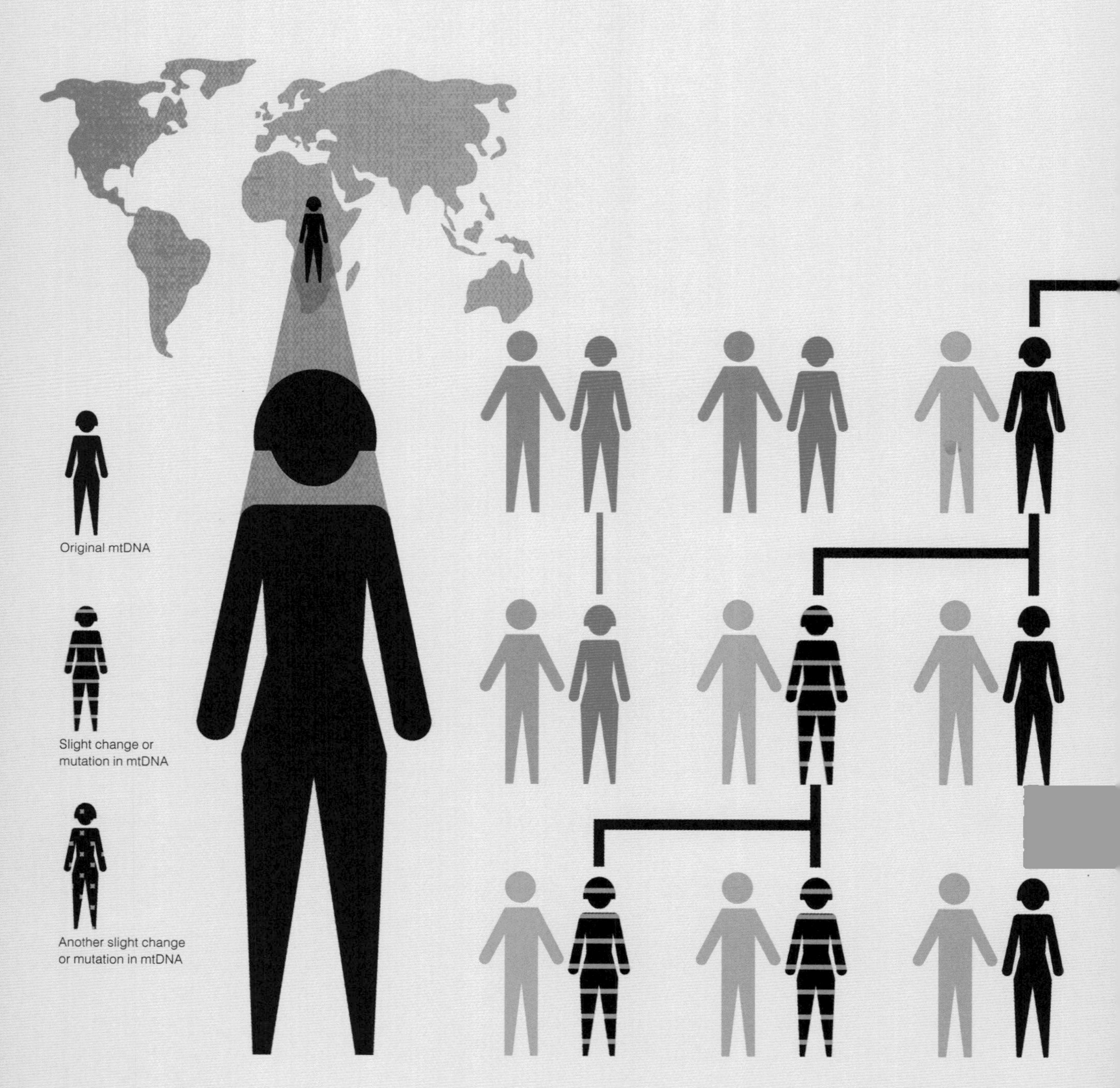

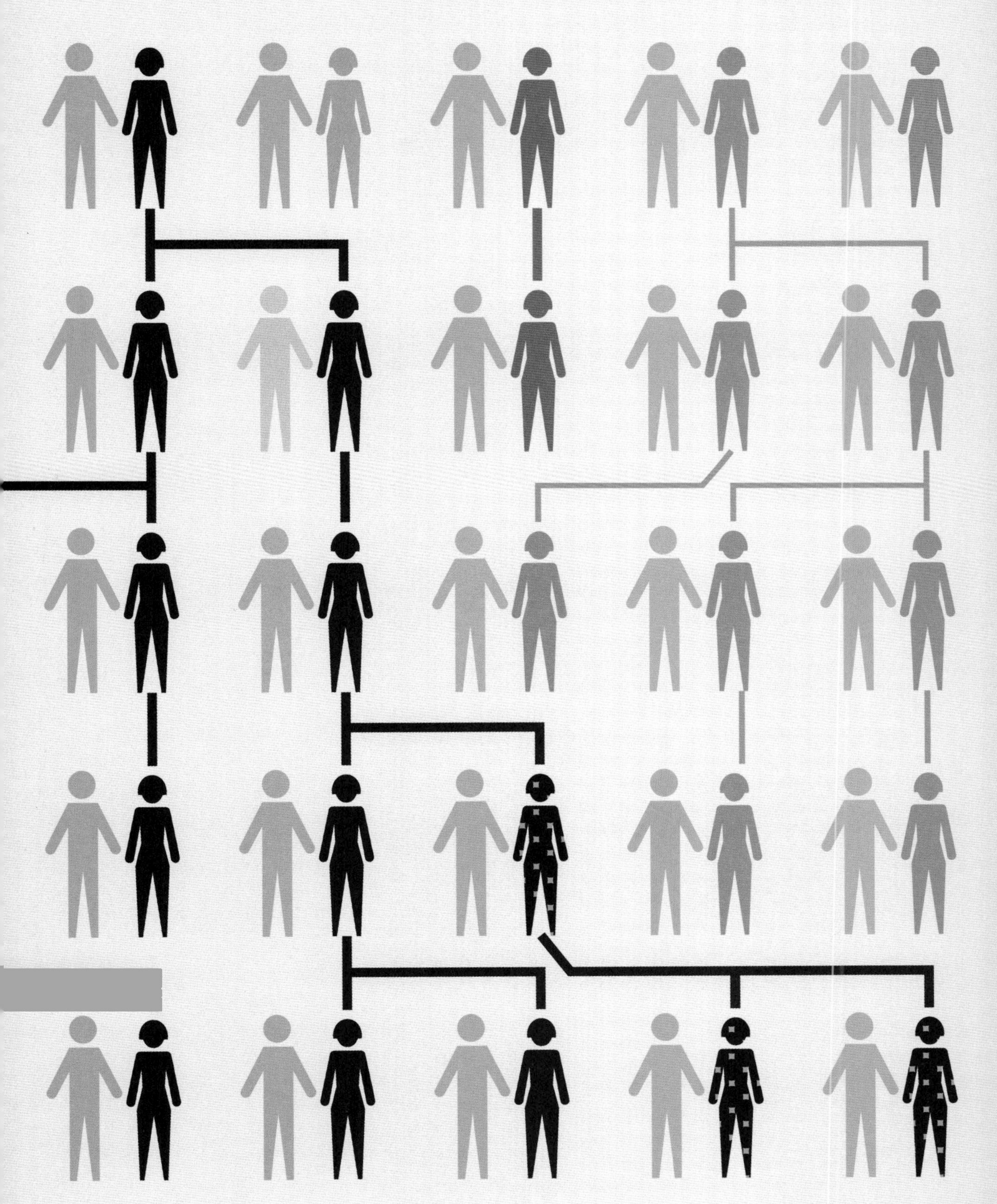

SENSITIVE BODY

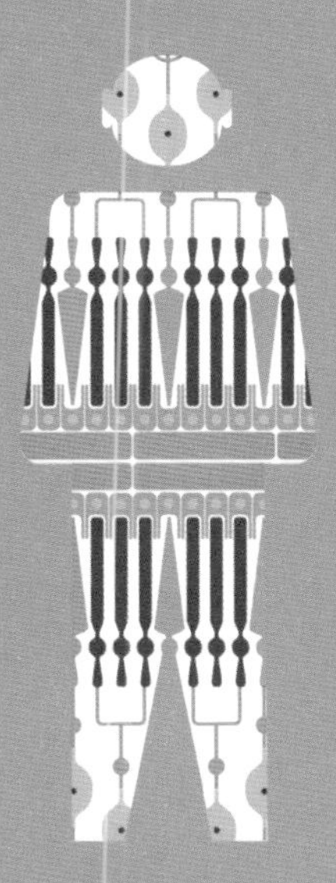

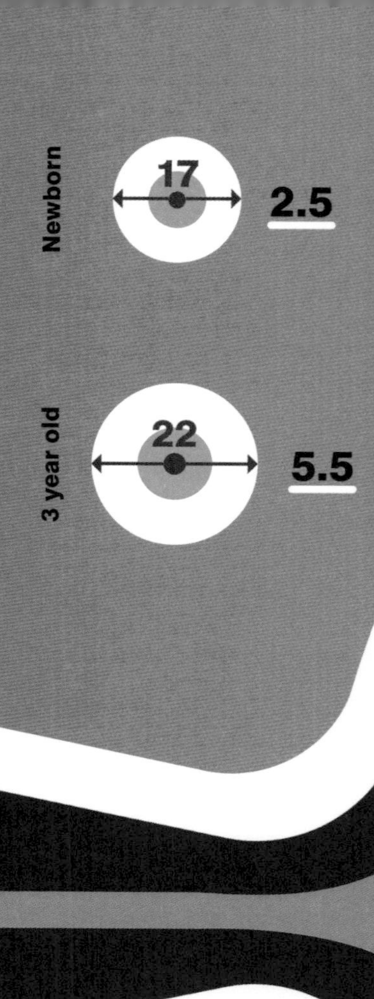

EYES HAVE IT

For sighted people, up to two-thirds of sensory information about the outside world enters through the eyes. Each of these real-time-roving, super-sharp-seeing, full-colour-capable living cameras is an intricate marvel of structures and tissues packed into a jelly-filled ball just 2.4 cm (1 inch) across. Light rays are bent, or refracted, through a series of almost perfectly transparent substances before they are detected by the retina, which then fires off nerve signals to the brain. To achieve such minimal light obstruction, the transparent tissues – cornea, lens, and aqueous and vitreous humours (liquids) – have the distinction of being the most blood-free in the body. By simple diffusion or seepage, the cornea gains nutrients from tear fluid and oxygen from air; the lens gets them from fluids around it.

Eyeball size

The eye is the organ nearest its final adult size at birth. Because of the way spheres enlarge, from newborn to adult its diameter increases by 41%, yet its volume enlarges by 188%.

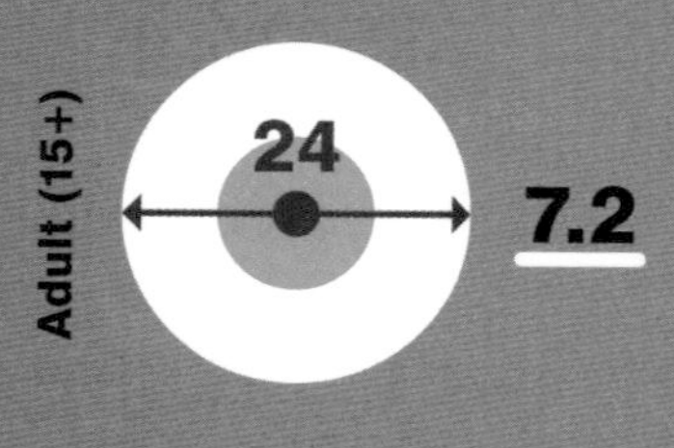

Diameter in mm

Volume in ml

VITREOUS

IRIS

Thickness ▼ (mm)

0.25 CONJUNCTIVA
Sensitive covering of eye, washed regularly by tear fluid and blinking

0.35 RETINA
Inner light-sensitive lining

0.5 CORNEA
Clear domed front of eye

Seconds to travel from an object 30 metres away

000,000,1

(one-tenth of one-millionth of a second)

1–1.5 AQUEOUS
Fluid between the cornea and lens, on either side of the iris

4 LENS
Elastic, adjusts to fine-focus light rays

PUPIL
Hole in centre of iris (Diameter)

The dimmer the light the wider the pupil

INSIDE THE RETINA

Our multi-colour, in-depth, continuously moving image of the world is sensed by an area not much larger than a thumbnail. The retina is packed with light-sensitive rod and cone cells, nerve fibres leading from them, a layer of nerve cells to network information from these fibres, another three layers of nerve cells for further information processing, and a branching web of blood vessels to supply all of these with oxygen and nourishment. A noticeable hurdle is that the cones and rods are almost at the base of the retina. So light must pass though all the other structures to get to them, causing plentiful obstructions and shadows. This could be viewed as a 'design fault', but the processing nerve cell layers and the brain itself rapidly become adept at computing what's likely to be there, to fill gaps.

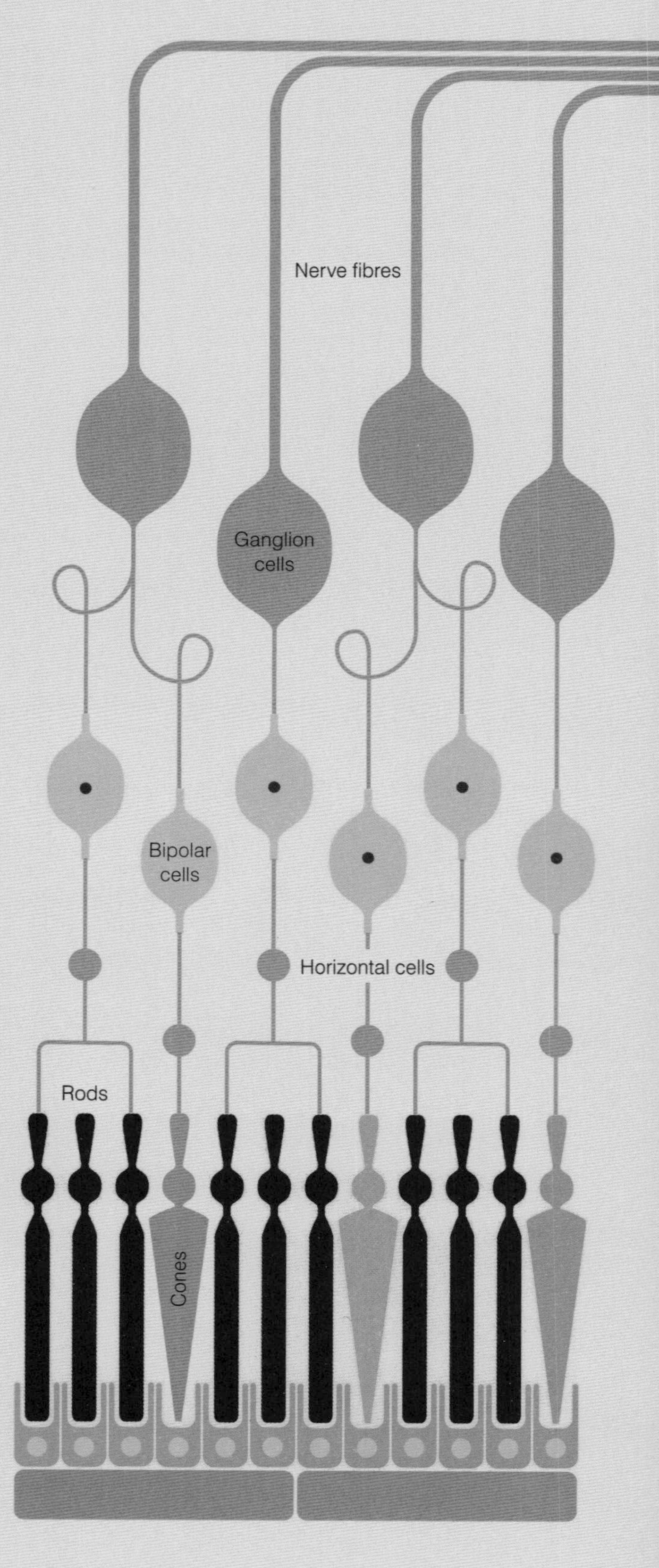

EYE VERSUS TV SCREEN

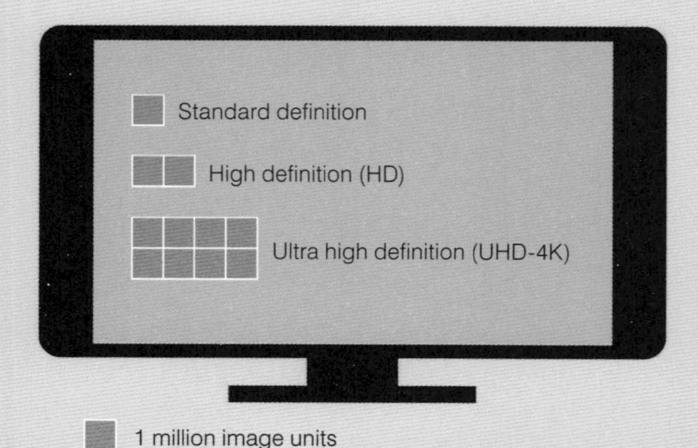

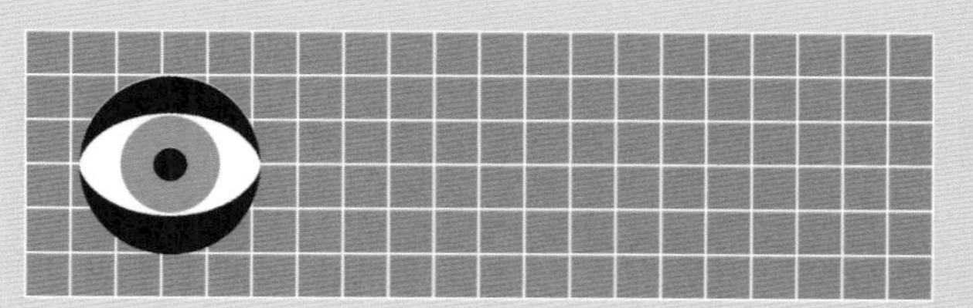

FIND YOUR BLIND SPOT!

Everybody has a blind spot, a point on the back of the retina where about one million ganglion cell nerve fibres gather to leave the retina as the optic nerve. There are no rods or cones here, so it corresponds to the 'blind spot'.

Close your right eye and look at the cross with your left eye. As you look at the cross move the page forwards and backwards until the eye blinks out of existence.

Try the same with this diagram. What happens to the black line when the cross disappears?

What happens when the eye is surrounded by colour?

What happens when the eye is surrounded by dots?

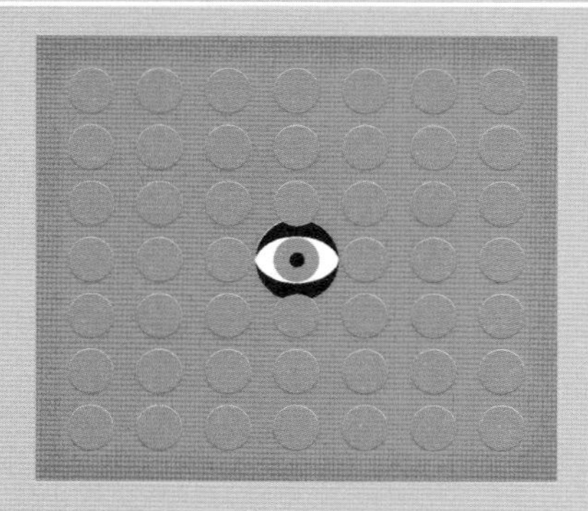

EYE TO BRAIN

What the eye sees is only part of what the mind's eye sees. We live in the past because of a gap, some 50–100 milliseconds (0.05–0.1 seconds), between rods and cones in the retina responding to light rays, and the mind's conscious perception of the image their nerve signals represent. Part of this delay is signals travelling through the networking cells of the retina, along the optic nerve, through the brain's optic chiasm (crossover) and nerve pathways to the main visual centres at the lower rear, then being 'shared around' to different accessory centres, each of which examines its own aspect of the scene. From all this information the mind constructs its own version of visual reality, looking backwards and forwards in time, analyzing and surmising, coordinating and correlating, always on the go but always lagging slightly behind.

FIELDS OF VISION

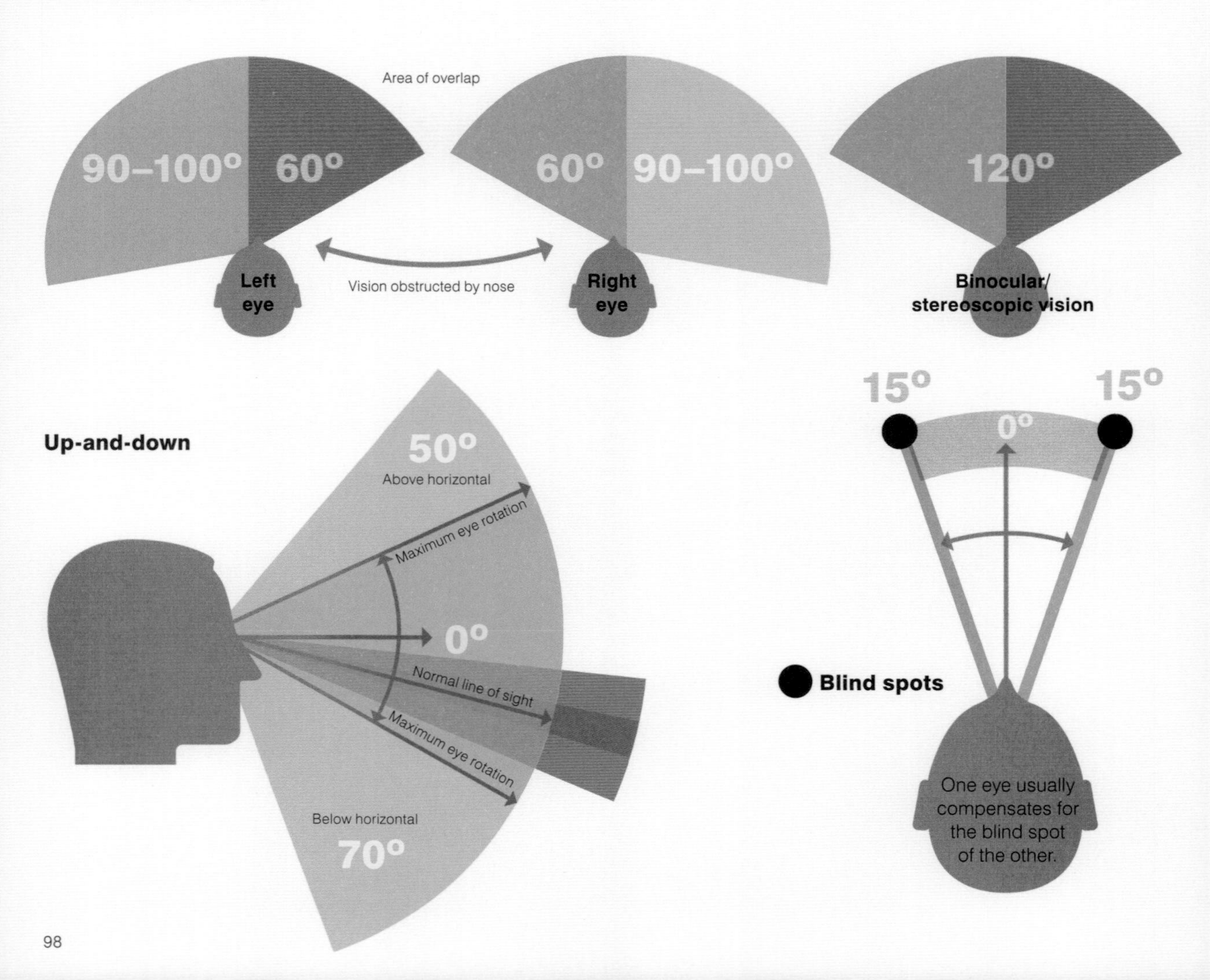

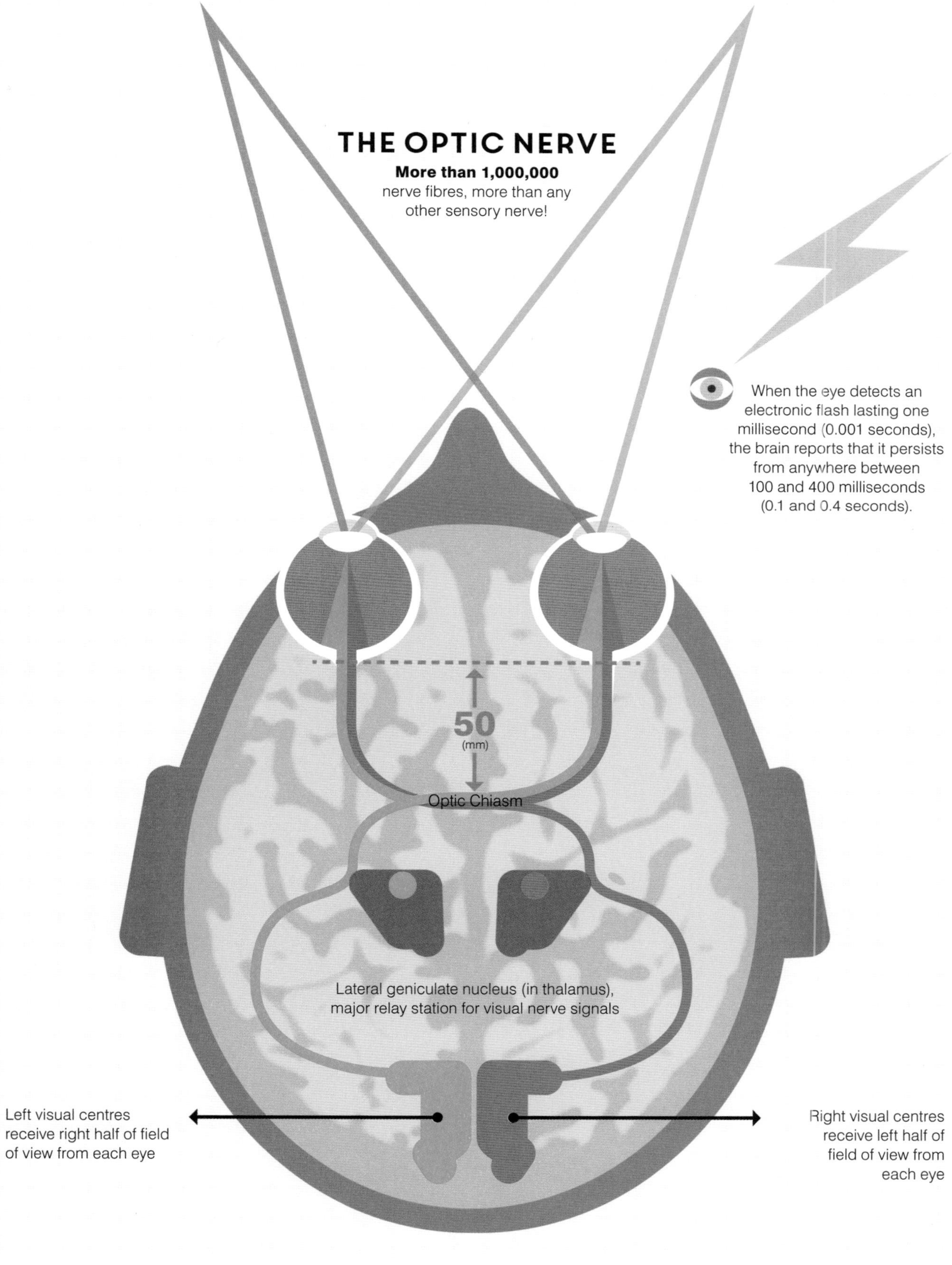
THE OPTIC NERVE
More than 1,000,000
nerve fibres, more than any
other sensory nerve!
When the eye detects an
electronic flash lasting one
millisecond (0.001 seconds),
the brain reports that it persists
from anywhere between
100 and 400 milliseconds
(0.1 and 0.4 seconds).
50
(mm)
Optic Chiasm
Lateral geniculate nucleus (in thalamus),
major relay station for visual nerve signals
Left visual centres
receive right half of field
of view from each eye
Right visual centres
receive left half of
field of view from
each eye

SOUND SENSE

A world awash with sound and noise comes from a little snail-shaped body part just 10 mm tall. It is deep in the inner ear but could comfortably perch on a little fingernail. The cochlea receives vibrations from the air via the eardrum and ear ossicle bones, and converts them into electrical nerve signals. Its key components are a row of about 3,500 inner hair cells arranged along a flexible sheet, the basilar membrane, winding inside the cochlea. As vibrations shake the sheet, the microhairs on the tops of these cells – which are embedded in a jelly-like 'roof' above – are bent and twisted. These ultra-fine motions cause the hair cells to generate nerve signals that zoom along the auditory nerve to the brain's hearing centres.

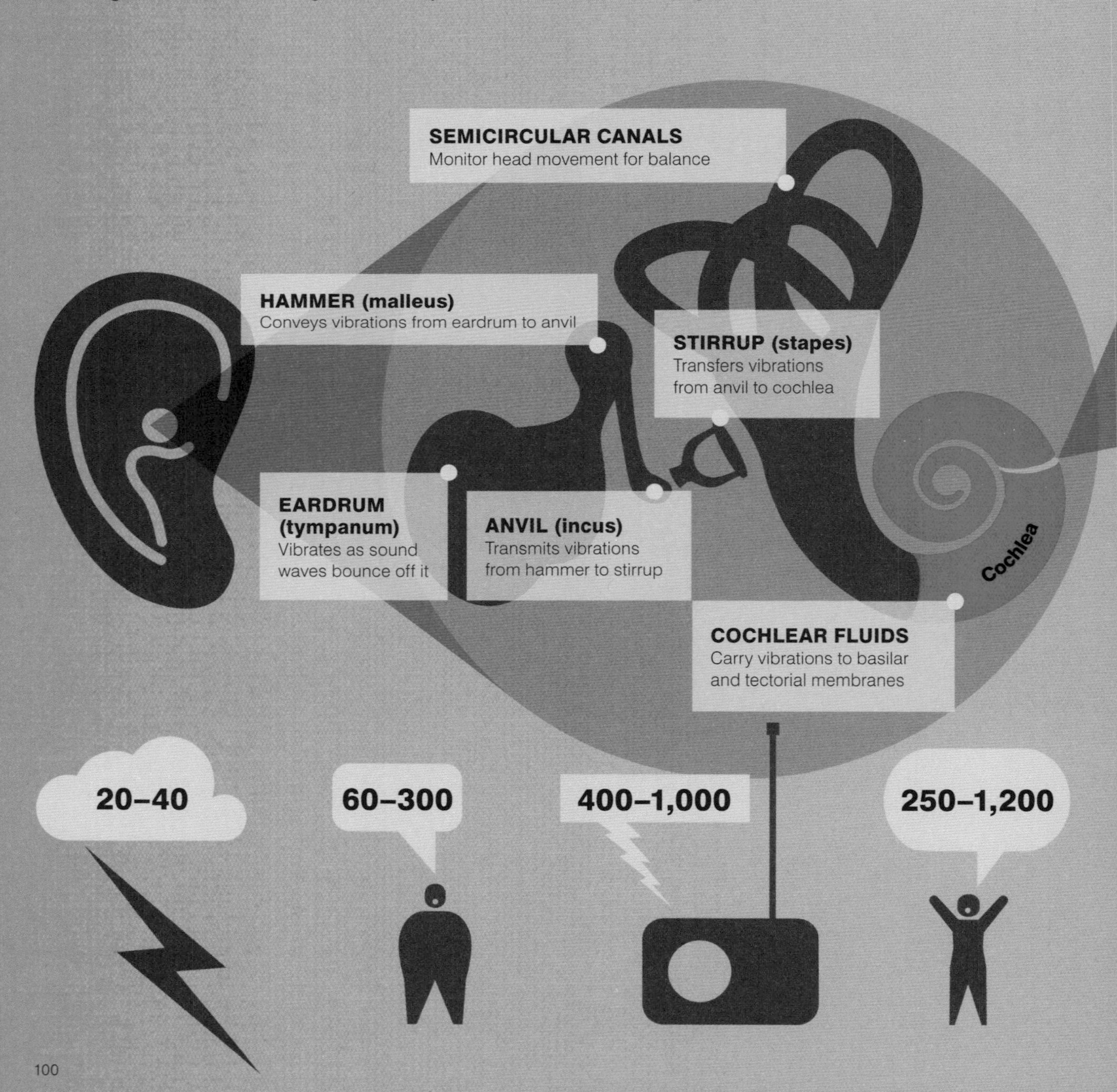

CROSS SECTION OF THE COCHLEA

Scala vestibuli

Scala media

Vibrations of membranes stimulate hair cells

Scala timpani

0.05 mm

12,000 OUTER HAIR CELLS
Receive nerve signals and move their microhairs to tense the basilar/tectorial membranes, increasing sensitivity of inner hair cells

0.03 mm

3,500 INNER HAIR CELLS
Vibrations cause microhairs to bend, producing nerve signals

HIGHS AND LOWS

Sound frequency or pitch is measured in vibrations per second, Hz (Hertz).

300–600

27.5

4,186

1,000–8,000

6,000

LIFE IN STEREO

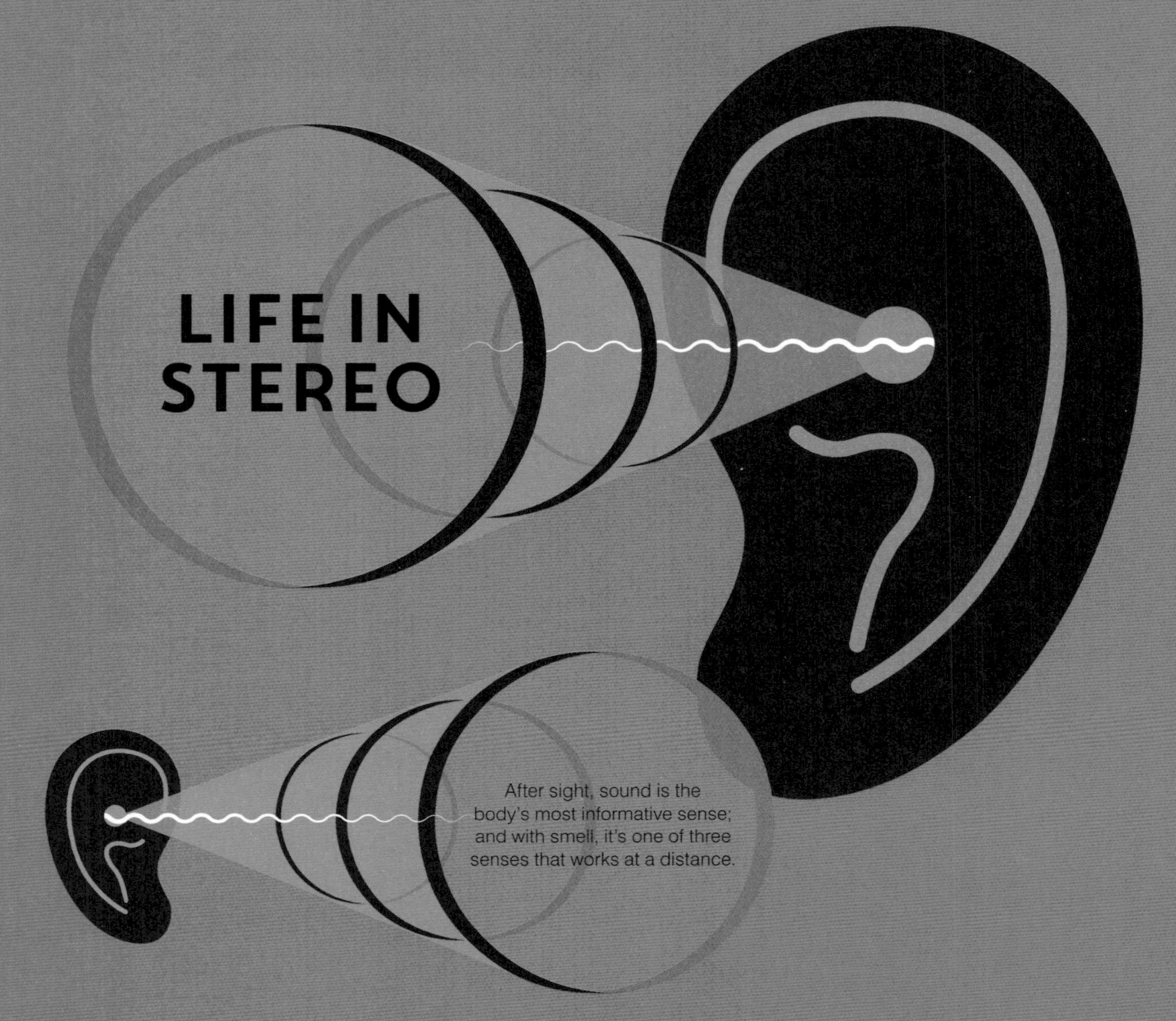

After sight, sound is the body's most informative sense; and with smell, it's one of three senses that works at a distance.

THE SPEED OF SOUND

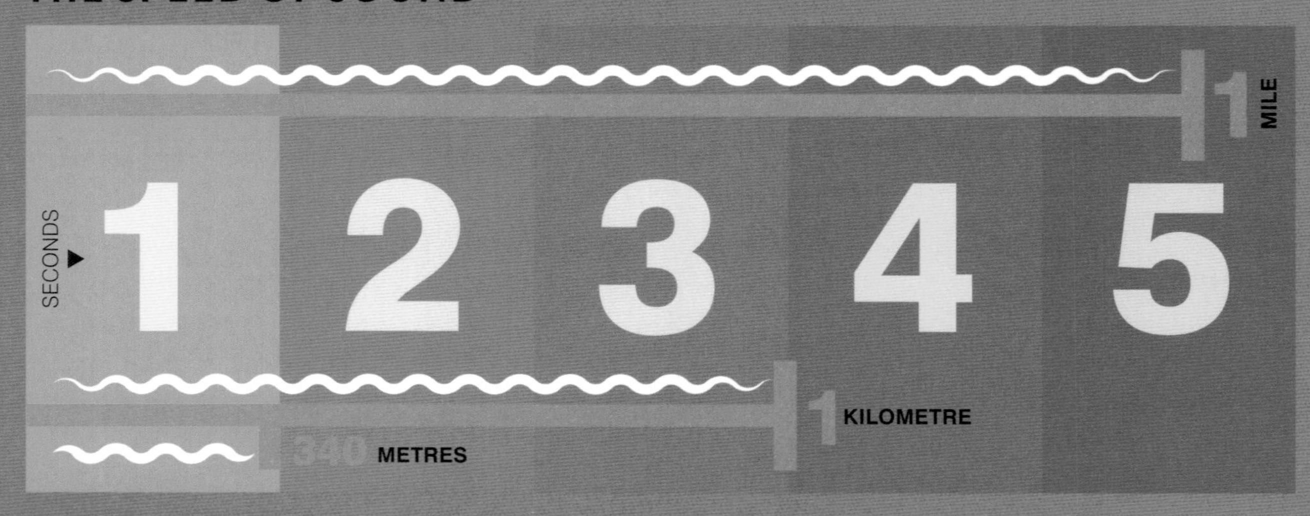

Sound speed is one million times slower than light speed, so ears use a delay system to gauge direction and distance. It's based on the gap between the ears, which means sounds from one side, such as music, reach the nearer ear less than 0.001 seconds before the far ear. These sounds are also quieter and more muffled in the far ear. Yet the brain's hearing centres detect all this in literally a split second. The brain then directs the neck muscles to turn the head to that side, to… face the music.

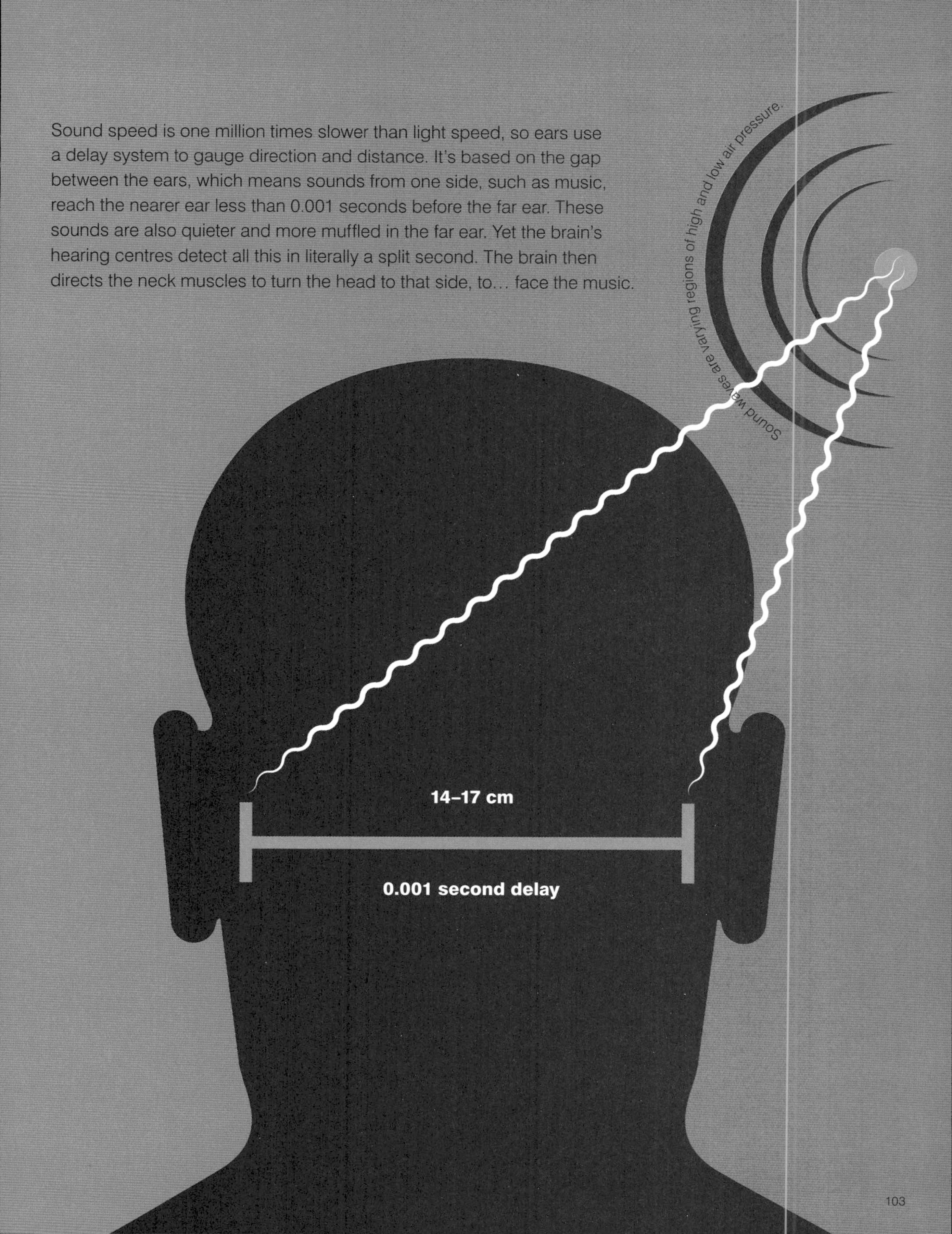

LOUD AND LOUDER

The sound intensity scale in decibels (dB) is not an equal step-wise progression scale but a logarithmic one to base 10. It means 20 dB is 10 times more intense or powerful than 10 dB (not twice), 30 dB is 100 times (not three), and so on.

170

Hearing loss inevitable

140

Jet engine at 30 m

120

Ear pain likely

110

Loud music concert, nearby thunder

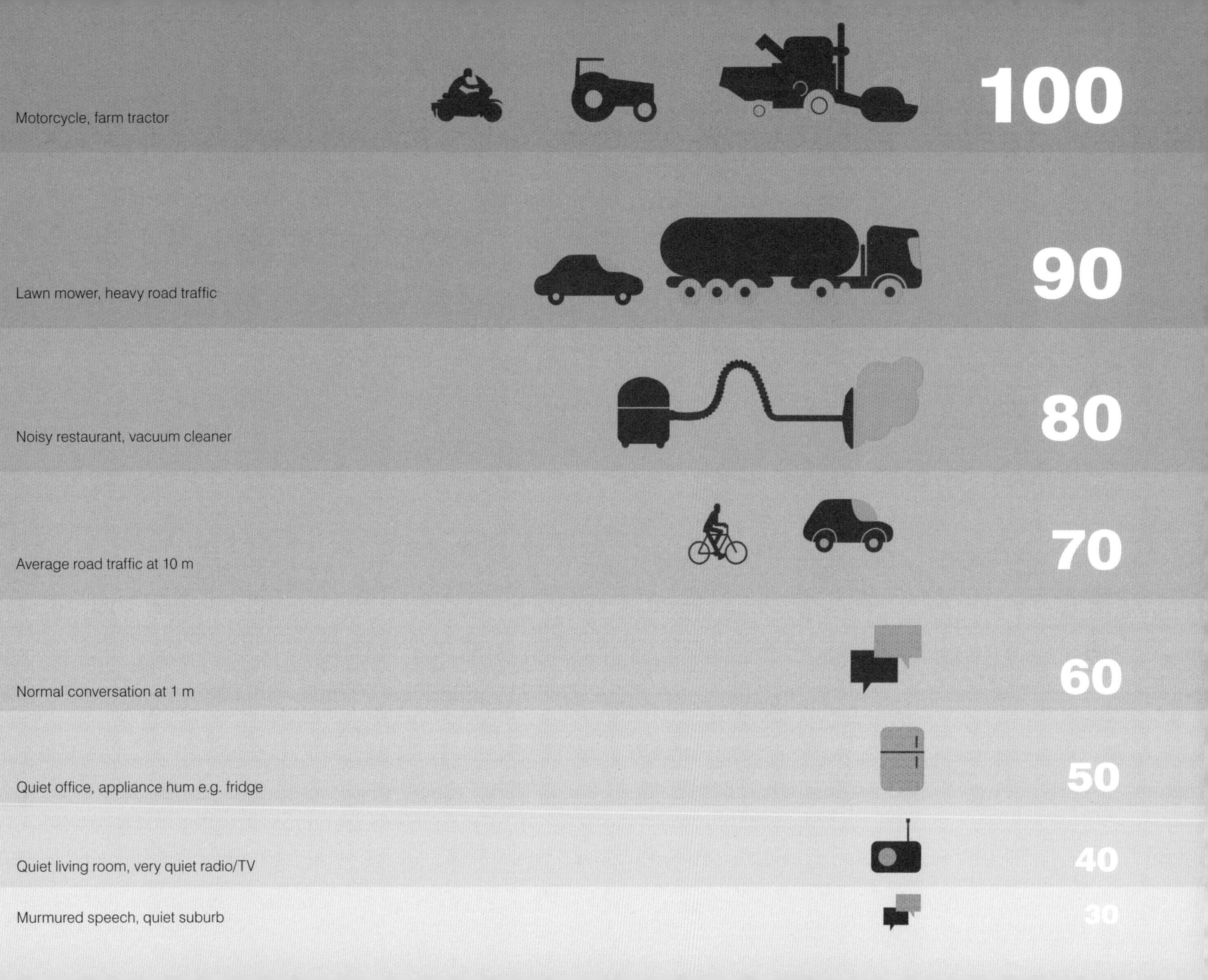

Motorcycle, farm tractor 100

Lawn mower, heavy road traffic 90

Noisy restaurant, vacuum cleaner 80

Average road traffic at 10 m 70

Normal conversation at 1 m 60

Quiet office, appliance hum e.g. fridge 50

Quiet living room, very quiet radio/TV 40

Murmured speech, quiet suburb 30

SCENTS SENSE

Smell or olfaction is the third most informative non-contact sense. It brings information about possibly dangerous vapours and gases in the air, as well as scents and aromas concerned with foods, drinks, plants, animals, and others of our kind – good and bad. Smells can give intense pleasure but also cause strong reactions, such as retching. More than other senses, olfaction is hard-wired to parts of the brain concerned with memories and emotions, which is why scents and aromas provoke such powerful feelings.

THE FOOD EXPERIENCE

Taste and smell are separate sensory systems but closely mingled in conscious perception to create the overall 'food experience' of each mouthful.

Estimated contribution to total 'food experience' (%):

15 Memories

15 Taste

60 Smell

10 Immediate circumstances

3

OLFACTORY EPITHELIUM
This area in the roof of the nasal chamber is 3 sq cm per side of nose and contains 5–10 million olfactory cells (olfactory receptor neurons). It produces fluid to dissolve odorant molecules and enable detection.

NASAL CHAMBER
Divided into two halves, left and right, by nasal septal cartilage. The lining warms, humidifies and filters particles from incoming air. Turbinates (conchae) are bony ridges directing airflow to olfactory epithelium.

2

1

ODORANTS
Invisible odour or smell particles (mostly molecules) float on air currents. They carry information in their size, shape and electrical charges, moving via the orthonasal route through nostrils from general surroundings and the retronasal route around rear of palate from food/drink in mouth.

5 NERVE FIBRES OF OLFACTORY CELLS
The nerve fibres collect into bundles of 20–30. They pass through cribriform plate, a perforated region of the ethmoid bone of the skull. They convey nerve signals to the olfactory bulb. They are regarded (sometimes with olfactory bulb and tract) as the olfactory nerve, also called cranial nerve I (1).

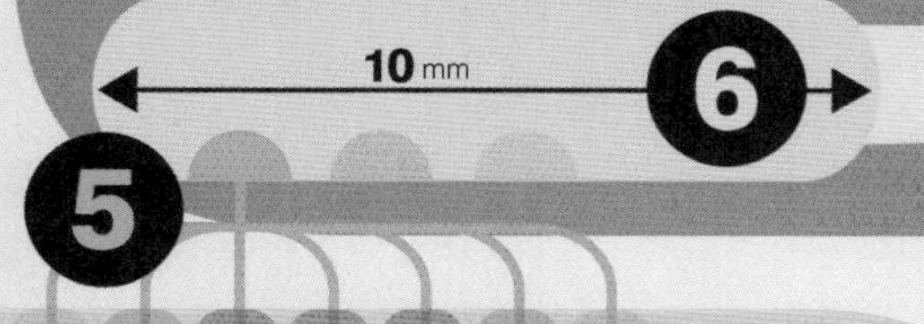

6 OLFACTORY BULB
This lobe-like extension of forebrain consists of five main cell layers. It decodes, filters, coordinates, enhances and processes nerve information from olfactory cells.

OLFACTORY RECEPTORS
These are molecules embedded in exposed surfaces of olfactory cells and they are stimulated on contact with a suitable odorant molecule – 'lock and key' mechanism. Olfactory receptor cells generate nerve signals, which are sent along the nerve fibres to the olfactory bulb.

7 OLFACTORY TRACT
This is the nerve fibres linking the olfactory bulb and brain.

8 PRIMARY OLFACTORY CORTEX
Main region dealing with smell information situated on the inner temporal lobe of the brain. It has close links with the regions involving emotions and memories.

25–30 mm

Volume (cc)
15–20

BEST POSSIBLE TASTE

In everyday experience, taste or gustation is inextricably mixed with smell, especially when savouring a fabulous meal. However, it is a separate sensory system – and itself is not all it seems. Nerve signals from its chief sensors, taste buds, provide only part of 'taste' information. Non-taste features such as heat/cold and physical texture – rough, slippery, squashy – add massively to the overall sensory impressions gained from foods. Researchers are finding that identifying tastes is similar in complexity to pinpointing smells. Many gustatory receptors fire many nerve signals at many rates when stimulated by many tastant substances (ones that stimulate the taste buds). The brain uses much decoding and pattern recognition to sort out the results. Bon appetit!

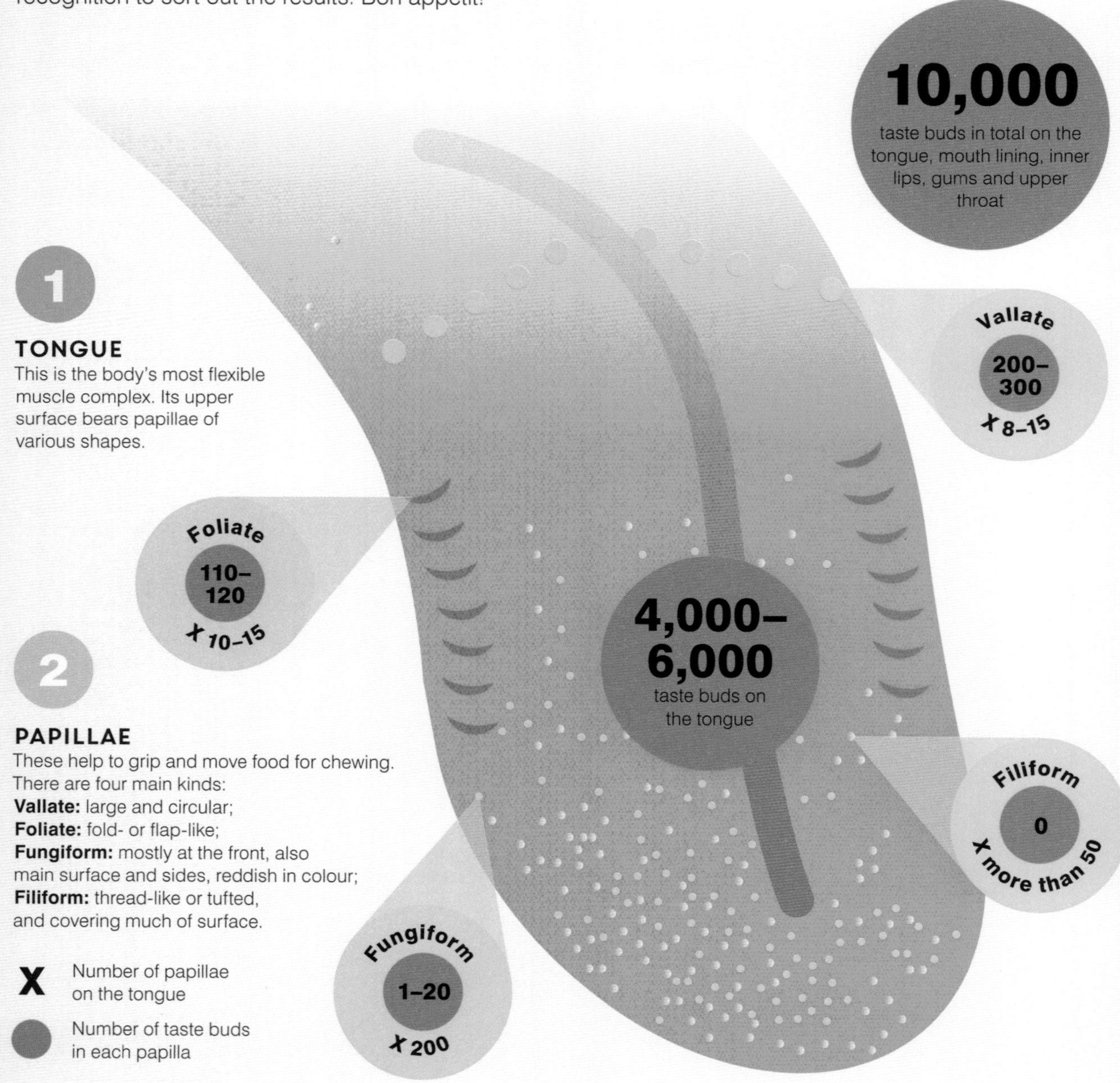

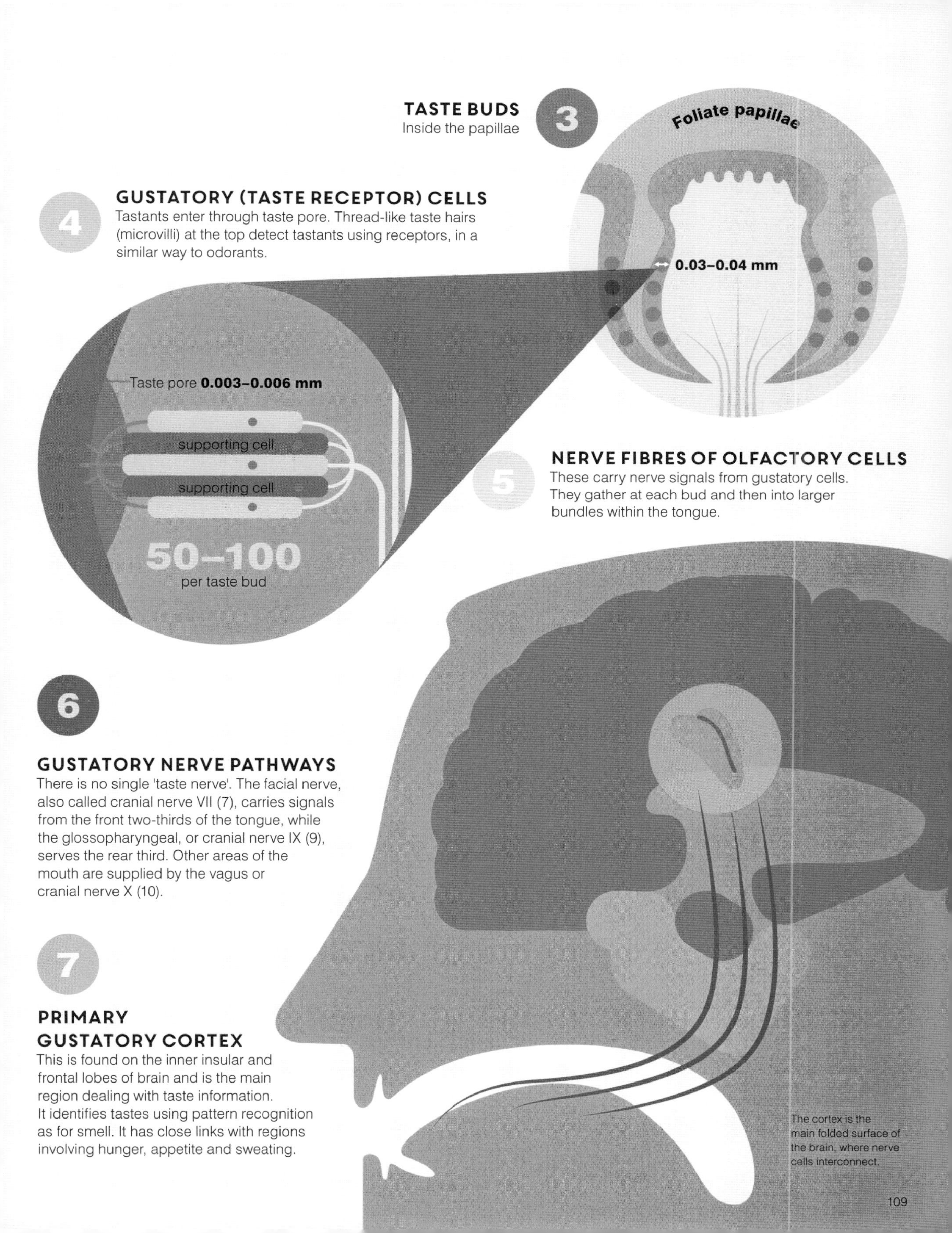
TASTE BUDS
Inside the papillae
3
Foliate papillae
4
GUSTATORY (TASTE RECEPTOR) CELLS
Tastants enter through taste pore. Thread-like taste hairs (microvilli) at the top detect tastants using receptors, in a similar way to odorants.
0.03–0.04 mm
Taste pore 0.003–0.006 mm
supporting cell
supporting cell
50–100
per taste bud
5
NERVE FIBRES OF OLFACTORY CELLS
These carry nerve signals from gustatory cells. They gather at each bud and then into larger bundles within the tongue.
6
GUSTATORY NERVE PATHWAYS
There is no single 'taste nerve'. The facial nerve, also called cranial nerve VII (7), carries signals from the front two-thirds of the tongue, while the glossopharyngeal, or cranial nerve IX (9), serves the rear third. Other areas of the mouth are supplied by the vagus or cranial nerve X (10).
7
PRIMARY GUSTATORY CORTEX
This is found on the inner insular and frontal lobes of brain and is the main region dealing with taste information. It identifies tastes using pattern recognition as for smell. It has close links with regions involving hunger, appetite and sweating.
The cortex is the main folded surface of the brain, where nerve cells interconnect.

SKIN AND SURFACE SENSORS

These are specialized nerve endings, regarded as single cells each giving off a nerve fibre.

20–100

Krause
Temperature changes
especially cold

Wilhelm Krause (German 1833–1910)

5–20

Merkel
Light touch, light pressure,
angular features such as edges

Friedrich Merkel (German 1845–1919)

100–300

Meissner
Light touch, slow vibrations,
surface textures

Georg Meissner (German 1829–1905)

TOUCHY FEELY

The skin's visible surface is actually dead cells designed for wear and protection, but just beneath teem millions of sensory cells. 'Touch' is far too simple a term. Types of contact distinguish rough or smooth, wet or dry, rigid or yielding, warm or cool, and many other features. These derive from floods of nerve signals from six main kinds of sensory cells. The signals pass through the bodywide nerve network to a strip on the brain's surface, the touch centre (officially the somatosensory cortex), where they register in conscious perception.

Sizes in μm (micrometres) 1 μm = 0.001 mm

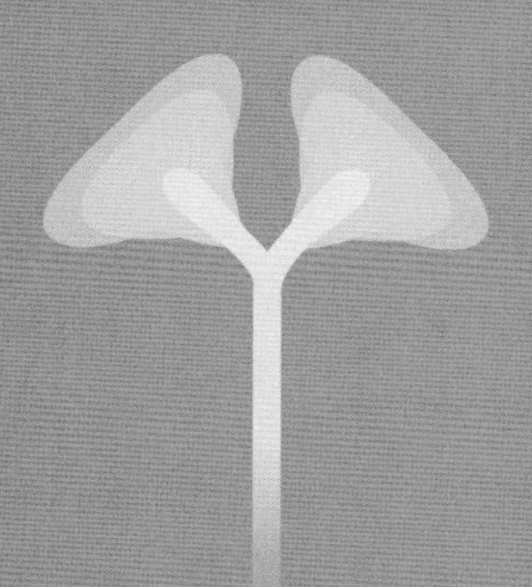

100–500

Ruffini

Slow-acting and sustained pressure, temperature changes especially heat

Angelo Ruffini (Italian 1864–1929)

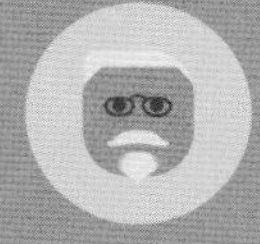

500–1,200

Pacini

Fast vibrations, strong pressure

Filippo Pacini (Italian 1812–1883)

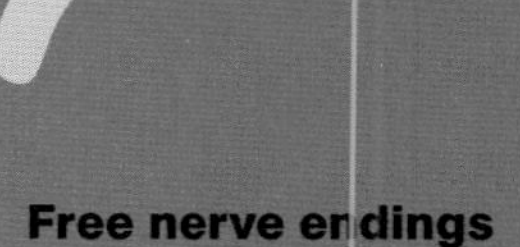

Free nerve endings

Various forms of touch, temperature changes, pain

Why the names?

Several skin sensors were named after the 19th-century anatomists, biologists or similar scientists who identified and studied them under the microscope.

INNER SENSE

Without looking, what are your arms and legs doing? Are they crossed, folded, straight, bent, stationary, moving? Knowing or being aware of the positions, postures and movements of body parts is termed proprioception. It is a sense we rarely consider but its second-by-second reports are vital in daily life. Its inputs are various tiny sensory organs and nerve endings that are mechanoreceptors (responding to physical forces). They are located almost everywhere within organs and tissues, especially in muscles and their tendons, and inside joint ligaments and capsules. Some are similar to those in the skin, such as Ruffini and Pacinci sensors (see page 111). As with the skin's touch messages, these proprioceptors send signals along nerves to the brain, where they are integrated with other sensory information to give an awareness of position and motion for each and every body part.

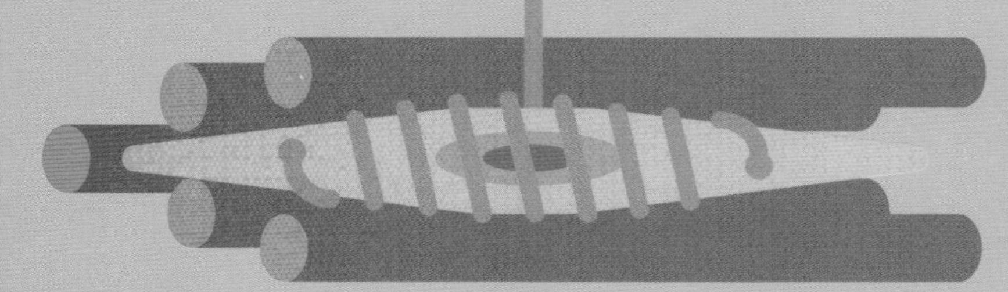

MUSCLE SPINDLES

Dozens to hundreds in main body or belly of a muscle. Respond to changes in length, detecting squeezing (compression) and stretching (tension)

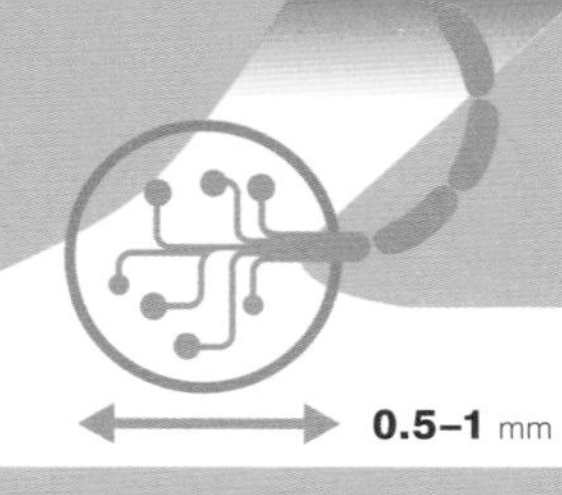

CAPSULE PROPRIOCEPTORS

In joint capsules, the fibrous casings around the bone ends in a joint. Similar to Ruffini and Pacini sensors in skin

0.1–1 mm

NEUROTENDON (GOLGI) SPINDLES

In tendons that connect muscles to bones. Respond to changes in squeezing (compression) as muscle contracts

LIGAMENT PROPRIOCEPTORS

0.1–1 mm

In stretchy ligaments that link bones in a joint. Similar to Ruffini and Pacini sensors in skin

PROPRIOCEPTIVE TESTS

Try these tests to demonstrate the importance of your inner sense. But note:

- First do the test quite quickly, without too much preparation or thinking.
- Then do it more slowly, concentrating on the positions of your arms and hands.
- With each further attempt, see how you can make your attention focus more precisely on proprioception.

1. Hold arms and hands straight to front.
2. Close eyes.
3. With left hand, touch tip of nose with thumb and each finger in turn.
4. Do same for right hand.

You will need:

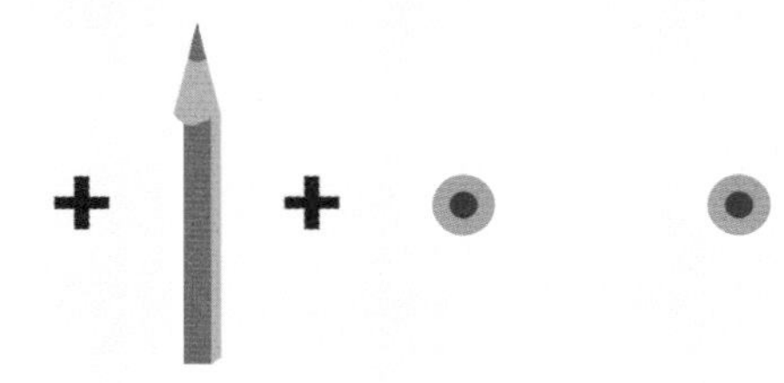

1. Sit at a table, and hold down a piece of paper with one hand.
2. Close eyes and keep them closed for duration of exercise.
3. With a pencil in your other hand, mark an X on the paper.
4. Swap the paper-holding and pencil-holding hands.
5. Mark a second X as near as possible to first X.
6. Open your eyes.

BALANCING ACT

Balance is sometimes characterized as a mysterious 'sixth sense'. In one respect, balance does engage the senses – in fact, almost all of the main sensory arrangements, plus others based deep in the ear. These inner-ear structures go by the general name of the vestibular system. It is based on the vestibule chamber of the inner ear with its offshoots of three semicircular canals, utricle and saccule. Their minutest detections, in parts known as the macula and ampulla, run along similar lines to hearing's cochlea, as hair cells fire off nerve signals when their microhairs are physically triggered. But balance is a much wider and continuously ongoing issue. It correlates constant inputs from eyes, skin and proprioceptive sensors, with continual outputs controlling muscles ranging from eye-movers to leg-stabilizers.

INNER EARS

Head movements cause fluid inside the ear to bend hair cells in cupulae of canals and maculae of vestibule chambers

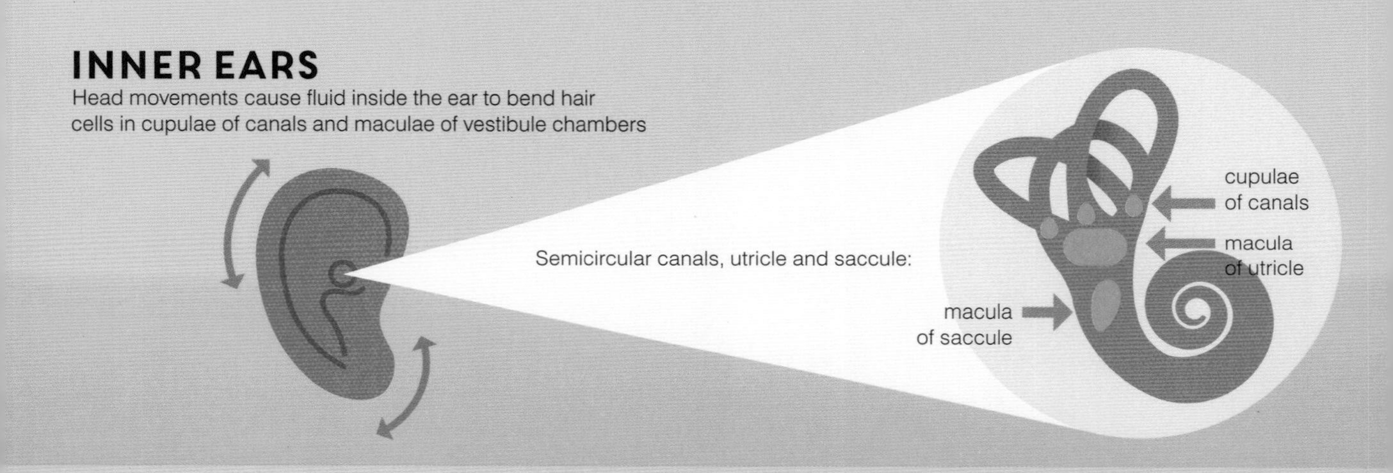

EYES

Register horizontals and verticals

PROPRIOCEPTORS

Pressure and tension sensors are found in:

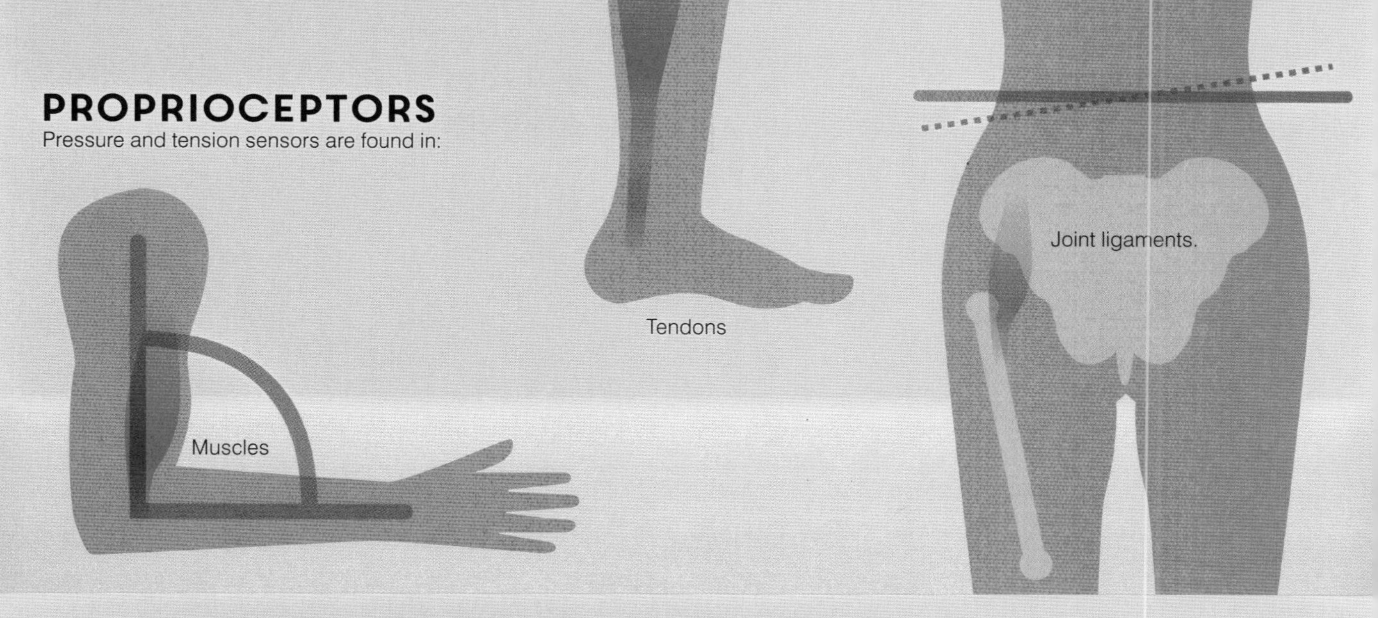

SKIN

Senses pressure, for example, pushing with hands and arms or the tilt on soles of the feet

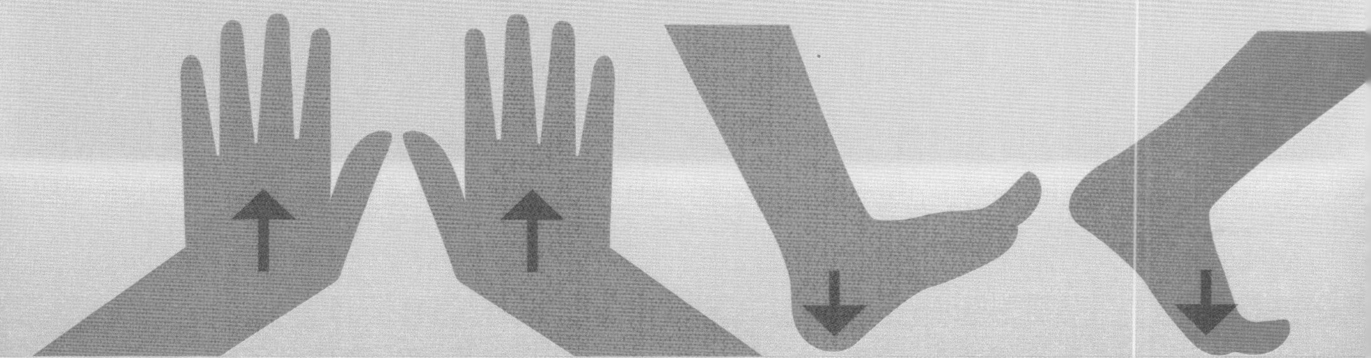

EARS

Detect incoming and reflected sounds

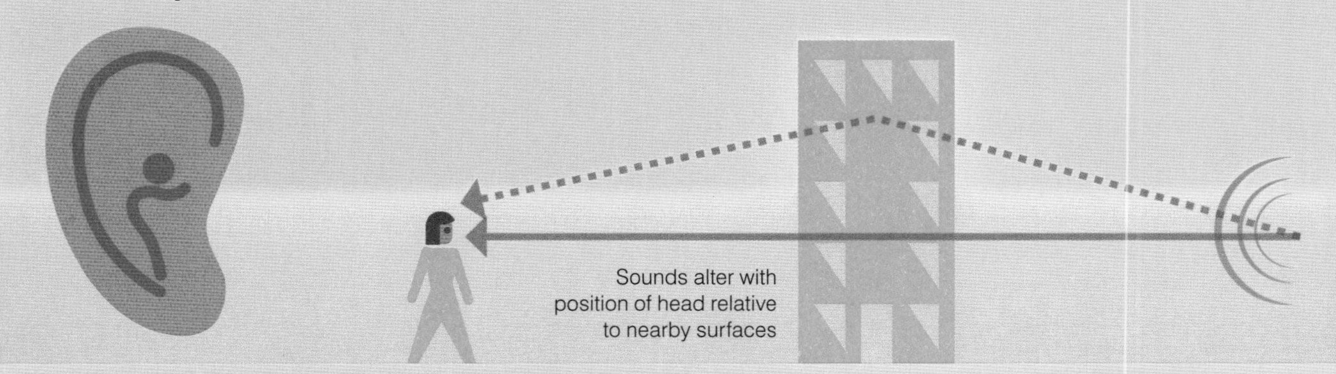

MAKING SENSE

Each main sense sends its nerve signals to its own region of the brain's thin, outer layer, the cortex. But on the way, these nerve signals and the information they represent go through many stages of processing, decoding, analyzing and being shared around. And once at the cortex, the information is likewise distributed and coordinated with other sensory centres for memories, recognition, naming, associations, emotions, decisions and reactions. Which is why an aroma familiar from childhood conjures up sights, sounds, tastes, feelings, even a whole recreation of a scene from long ago. Pine woodland, seaside spray, theme park snack, regurgitated baby milk ...

Lobes of the brain

The lobes of the main brain parts, the cerebral hemispheres, are anatomical areas known since antiquity. They are demarcated by deep grooves called fissures or sulci.

FRONTAL	• conscious thought • self-awareness • decisions • personality • memories • aspects of smell and speech • planning and controlling movements
CENTRAL SULCUS	Separates frontal and parietal lobes
PARIETAL	• coordinating sensory information • visual-spatial aspects • various aspects of touch • taste • aspects of speech • proprioception
LATERAL SULCUS (FISSURE)	Separates frontal and parietal lobes from temporal lobe
LIMBIC	• emotions • memories • experiences
PARIETO-OCCIPITAL SULCUS	Separates parietal and occipital lobes
OCCIPITAL	• sight and associated, features • sensory coordination • memories
TEMPORAL	• hearing • aspects of smell and sight • coordination of sensory information • speech • language • short- and long-term memories

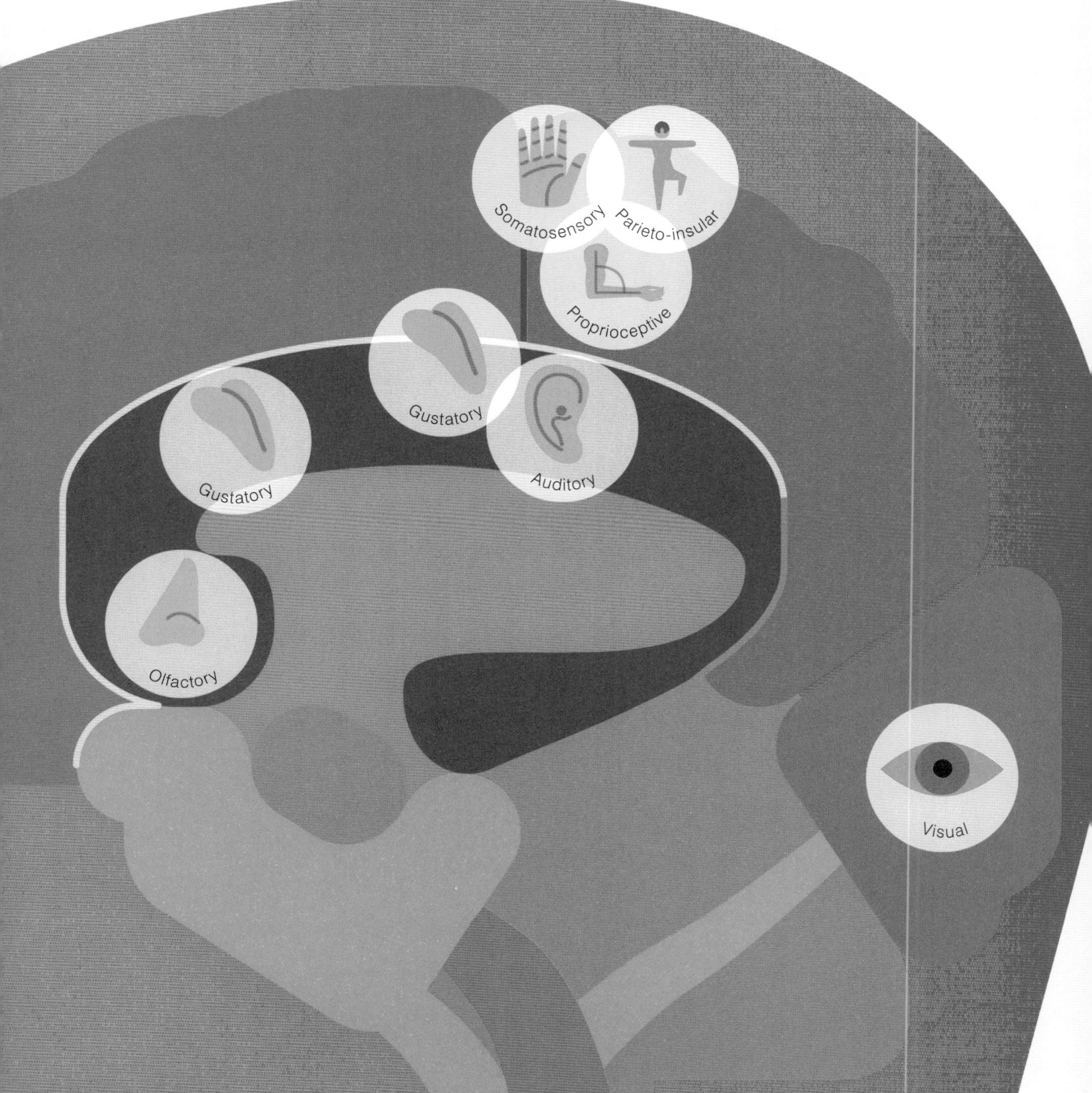

Sensitive brain
The main areas for receiving sensory information are each demarcated on a particular lobe. Oddly, the surface of the brain itself has no touch sensory cells and so, if poked or prodded, feels no contact (although consciousness may be affected).

TOUCH MAP

The somatosensory cortex (touch centres) on each side of the brain form a strip-like body map. More sensitive parts like lips and fingertips occupy more area on the cortex.

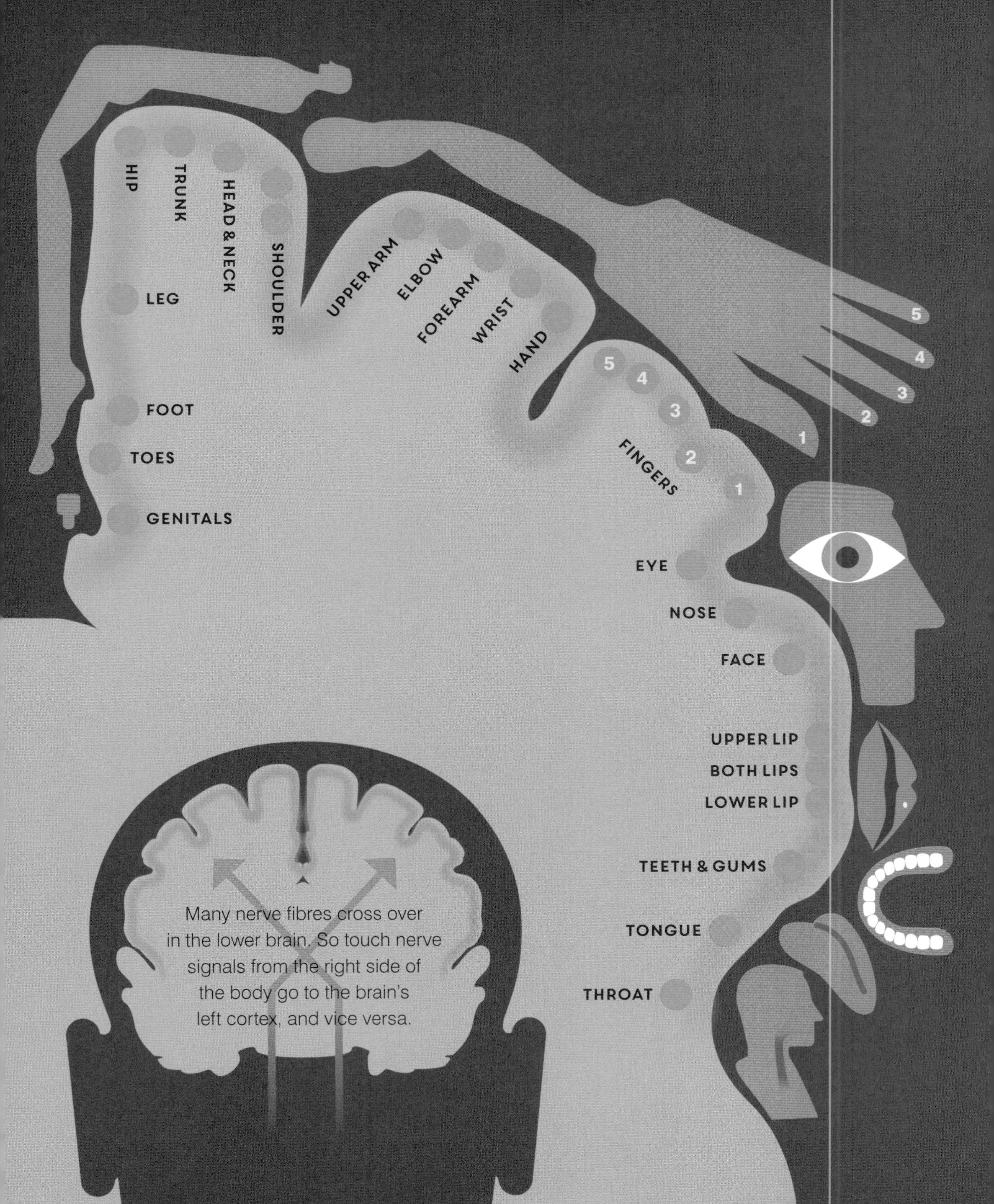
HIP
TRUNK
HEAD & NECK
SHOULDER
UPPER ARM
ELBOW
FOREARM
WRIST
HAND
LEG
FOOT
TOES
GENITALS
5
4
3
2
1
FINGERS
EYE
NOSE
FACE
UPPER LIP
BOTH LIPS
LOWER LIP
TEETH & GUMS
TONGUE
THROAT
Many nerve fibres cross over in the lower brain. So touch nerve signals from the right side of the body go to the brain's left cortex, and vice versa.

COORDINATED BODY

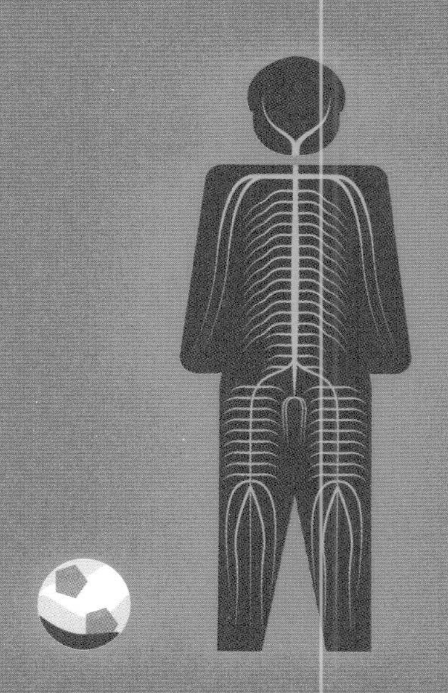

FEELING NERVOUS

The body's billions of cells, hundreds of tissues and dozens of organs work together as a harmonious whole – but how? There are two chief bodywide coordination-control-command systems: nervous, and hormonal or endocrine. The first works mainly by tiny electrical signals whizzing along wire-like nerves, while the second is based on chemical substances called hormones. The brain is the hub of both systems.

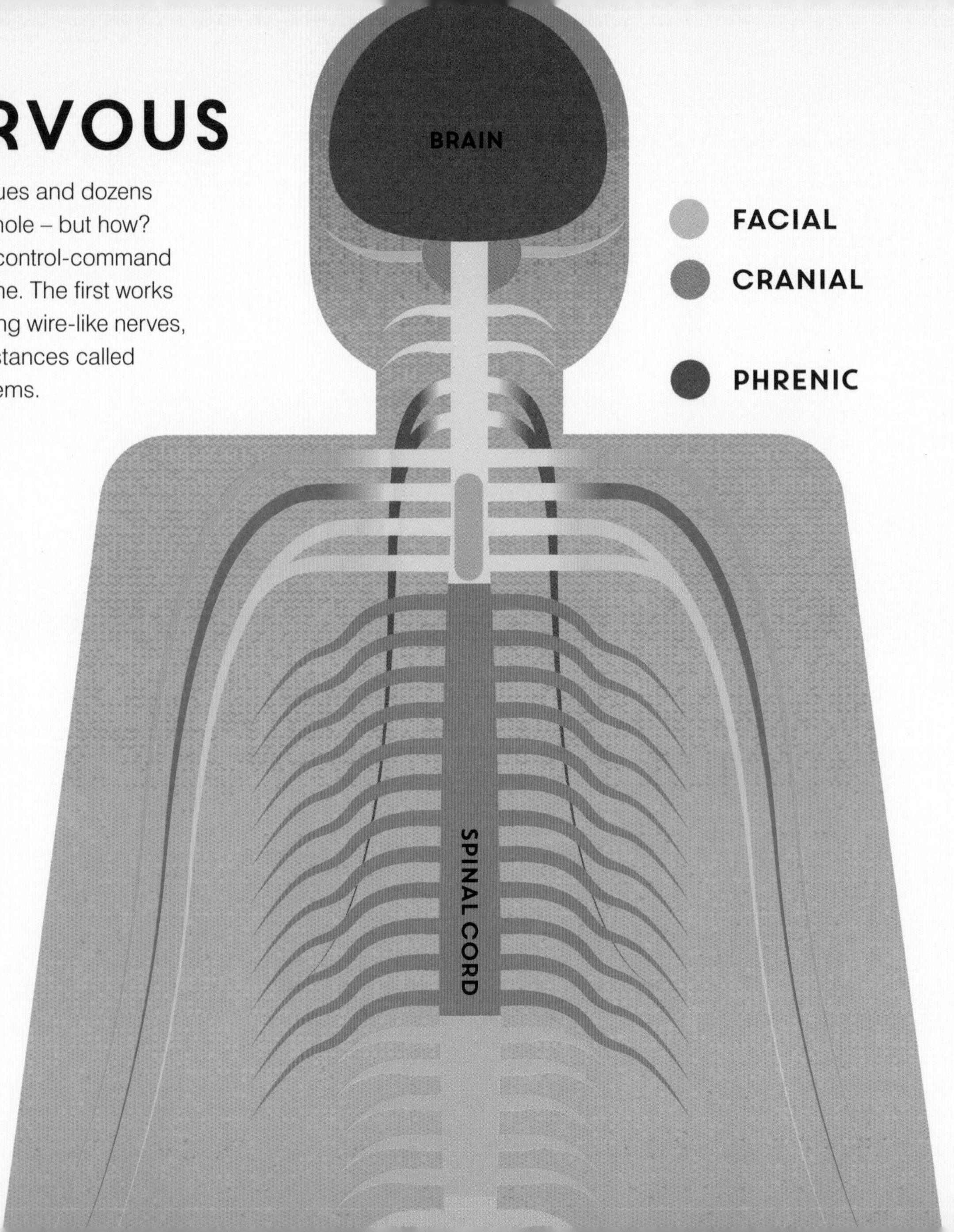

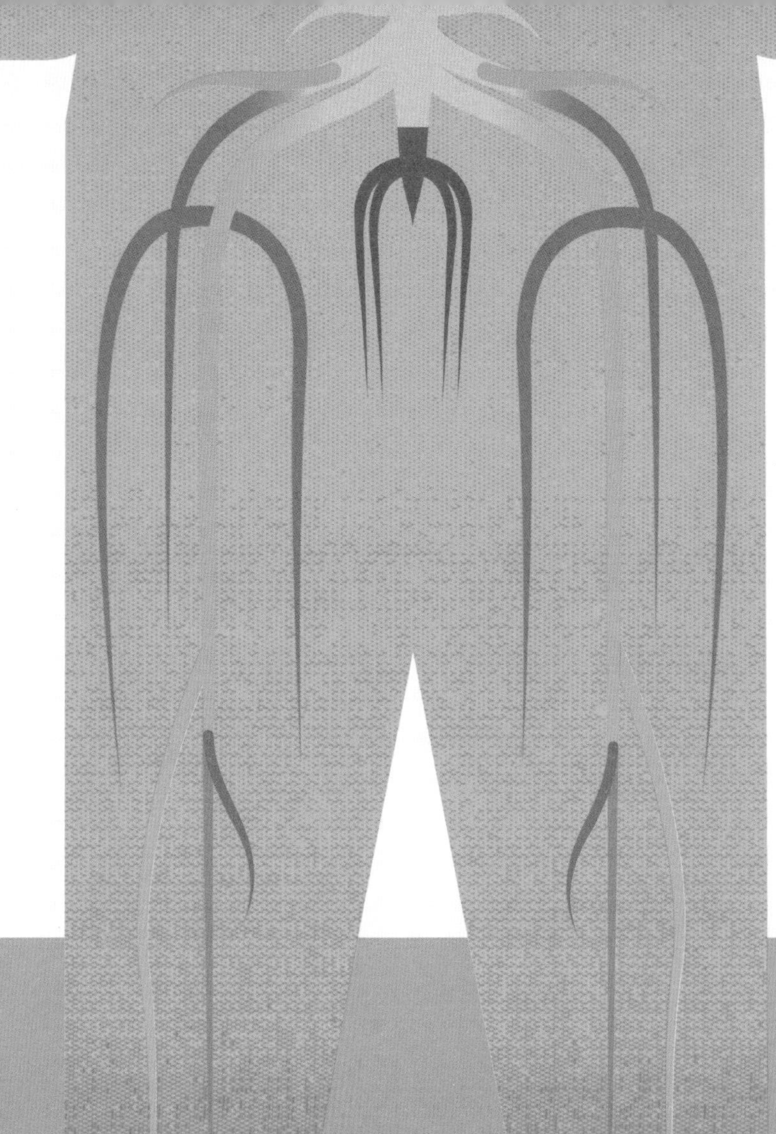

NERVE MAP

Nerves branch from the brain or spinal cord, dividing repeatedly as they connect into every body part and become microscopically thin.

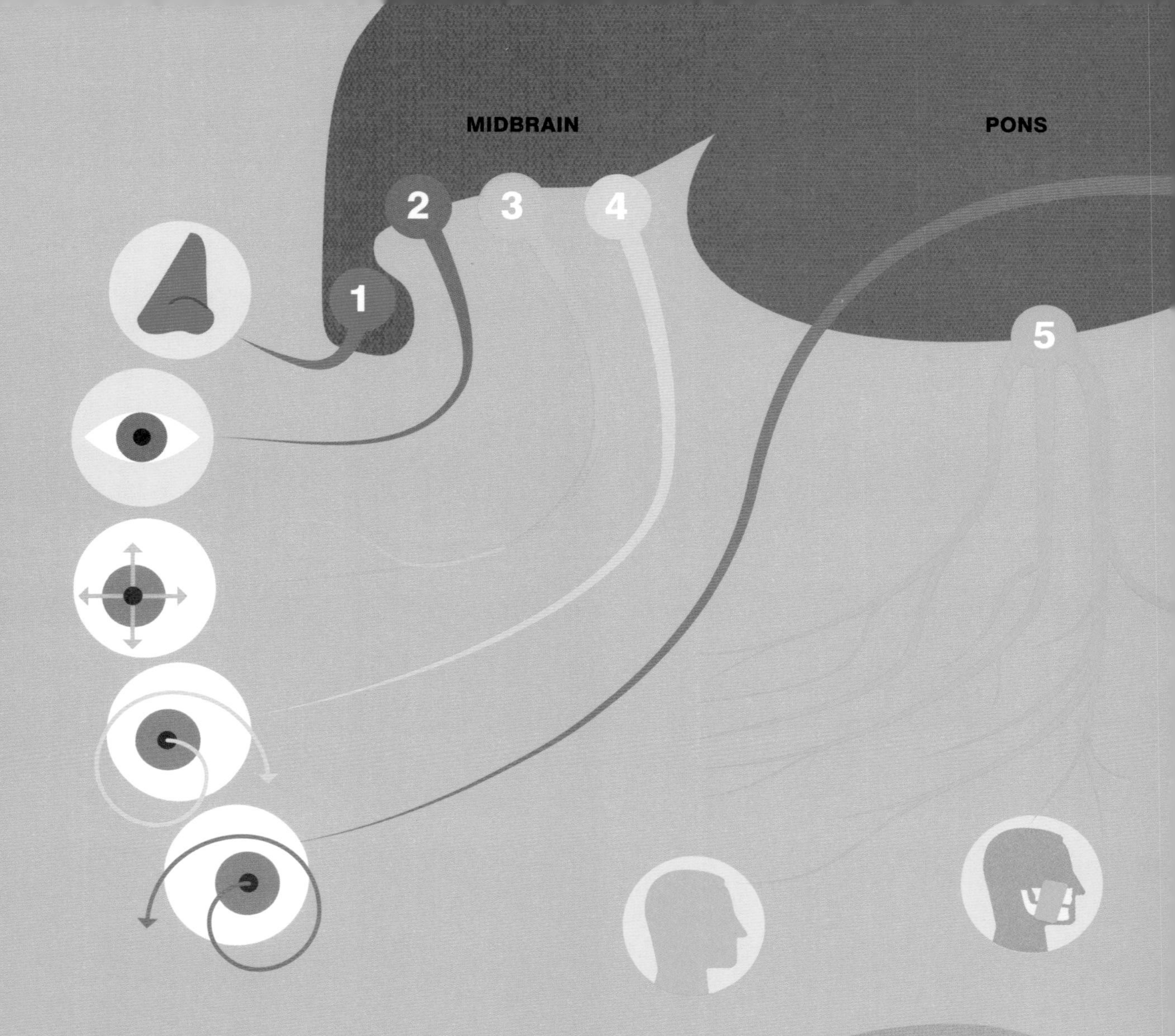

HEADFUL OF NERVES

Forty-three pairs of nerves (left and right) branch from the brain and spinal cord out into the body. Twelve of these come directly from the brain and are known as cranial nerves; the other 31 pairs from the cord are spinal nerves. Cranial nerves carry information from the main senses to the brain, and signals from the brain to the muscles of the face, head and neck – and in one case, the heart, lungs and stomach.

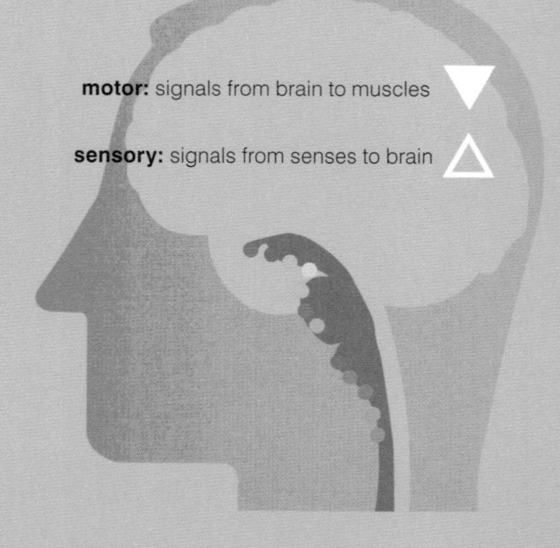

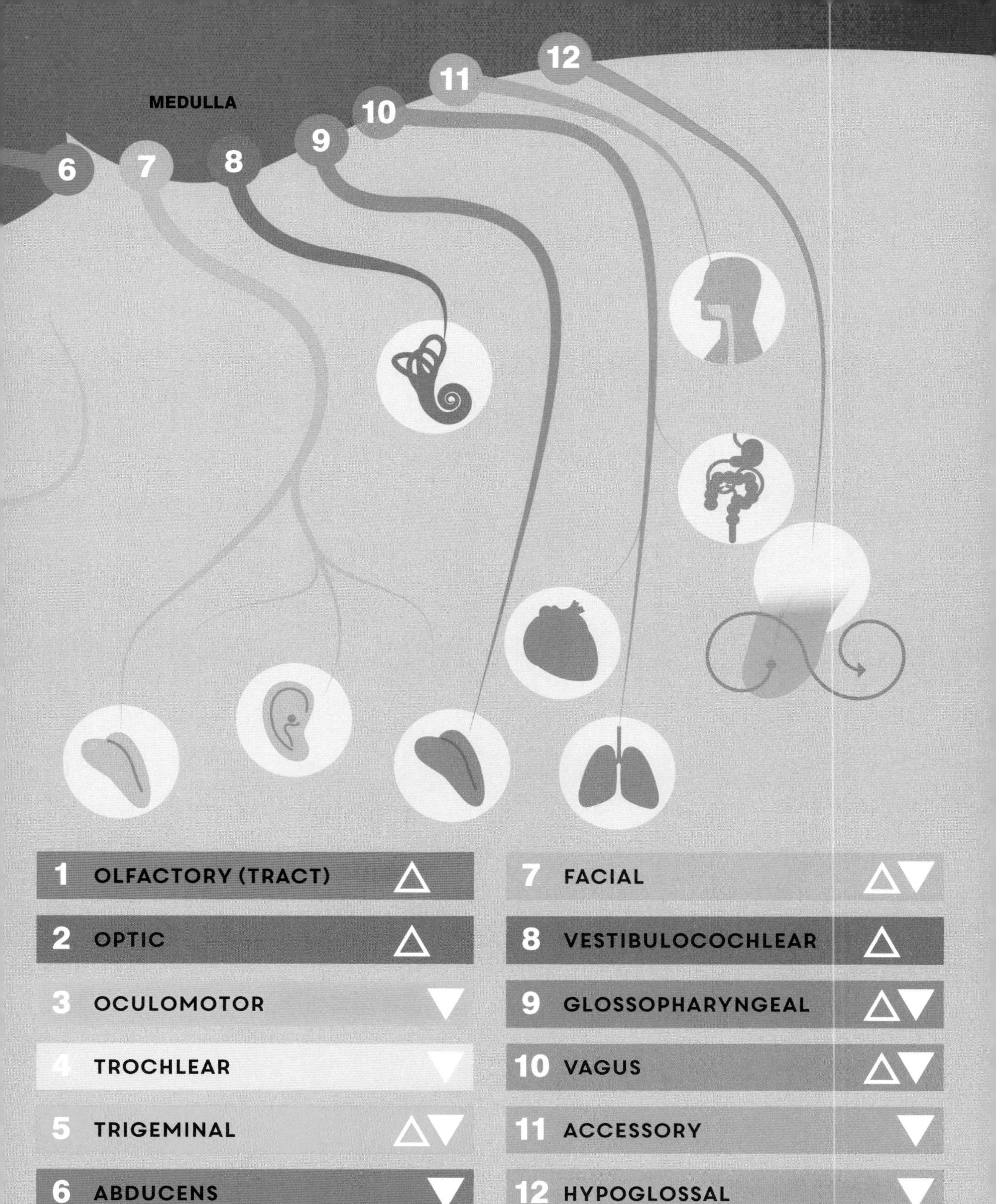
MEDULLA
6
7
8
9
10
11
12
1 OLFACTORY (TRACT)
2 OPTIC
3 OCULOMOTOR
4 TROCHLEAR
5 TRIGEMINAL
6 ABDUCENS
7 FACIAL
8 VESTIBULOCOCHLEAR
9 GLOSSOPHARYNGEAL
10 VAGUS
11 ACCESSORY
12 HYPOGLOSSAL

MIND THE GAP

Nerves all around the body use the same basic communication system. This is chiefly electrical with some chemical stages along the way. A single nerve signal or message is a tiny pulse of electricity that lasts for a tiny amount of time; it is the same for any nerve, at any time, anywhere in the body. The information carried depends on how fast the pulses follow each other, where they come from, and where they go.

1 INCOMING

Nerve signals are gathered by nerve cell's dendrites

Dendrite size range
0.1–5 μm

2 SIGNAL

Also called an action potential, this is caused by electrically charged particles (ions) moving across the cell membrane.

0.1VOLTS
1MILLISECOND

3 INTEGRATION

Nerve cell (neuron) receives perhaps millions of signals per second. Some signals reinforce many other interactions, while others cancel them.

Range of nerve cell body size
5–50 μm

4 OUTGOING

Resulting signals leave cell body along axons (nerve fibres)

Diameter range of axons
0.2–20 μm

Some nerve cells have over
10,000
dendrites totalling several centimetres in length.

5
IMPROVING CONDUCTION

Myelin sheath coats many axons. The sheath of fatty myelin spirals around the axons. This increases speed as signal 'jumps' along the axon. The sheath also prevents the signal from weakening and reduces leakage.

6
AT THE JUNCTION

The junction between nerve cells is called a synapse. The end of each axon does not quite touch the next nerve cell.

Average synaptic gap
0.02 μm

8
ONWARDS

The receiver is another nerve cell's dendrite or cell body. Neurotransmitters trigger new electrical signals. Signals travel away from the synapse.

0.1 MILLISECOND
Crossing time

7
CHEMICAL CROSSING

Chemicals called neurotransmitters carry the signal. Each signal can have thousands, even millions of neurotransmitter molecules.

Longest axons almost 1m (toe to spinal cord)

1 μm = micrometre = 0.001 mm = 0.000,001 m (one millionth of a metre)

THE VITAL LINK

The spinal cord is the brain's long, slim, metro-like link into the main body. Thirty-one pairs of spinal nerves branch from it, exiting at joints between the backbones or vertebrae. All the spinal nerves carry sensory information from the skin and inner organs via the spinal cord to the brain, and motor signals from the brain via the spinal cord out to the muscles.

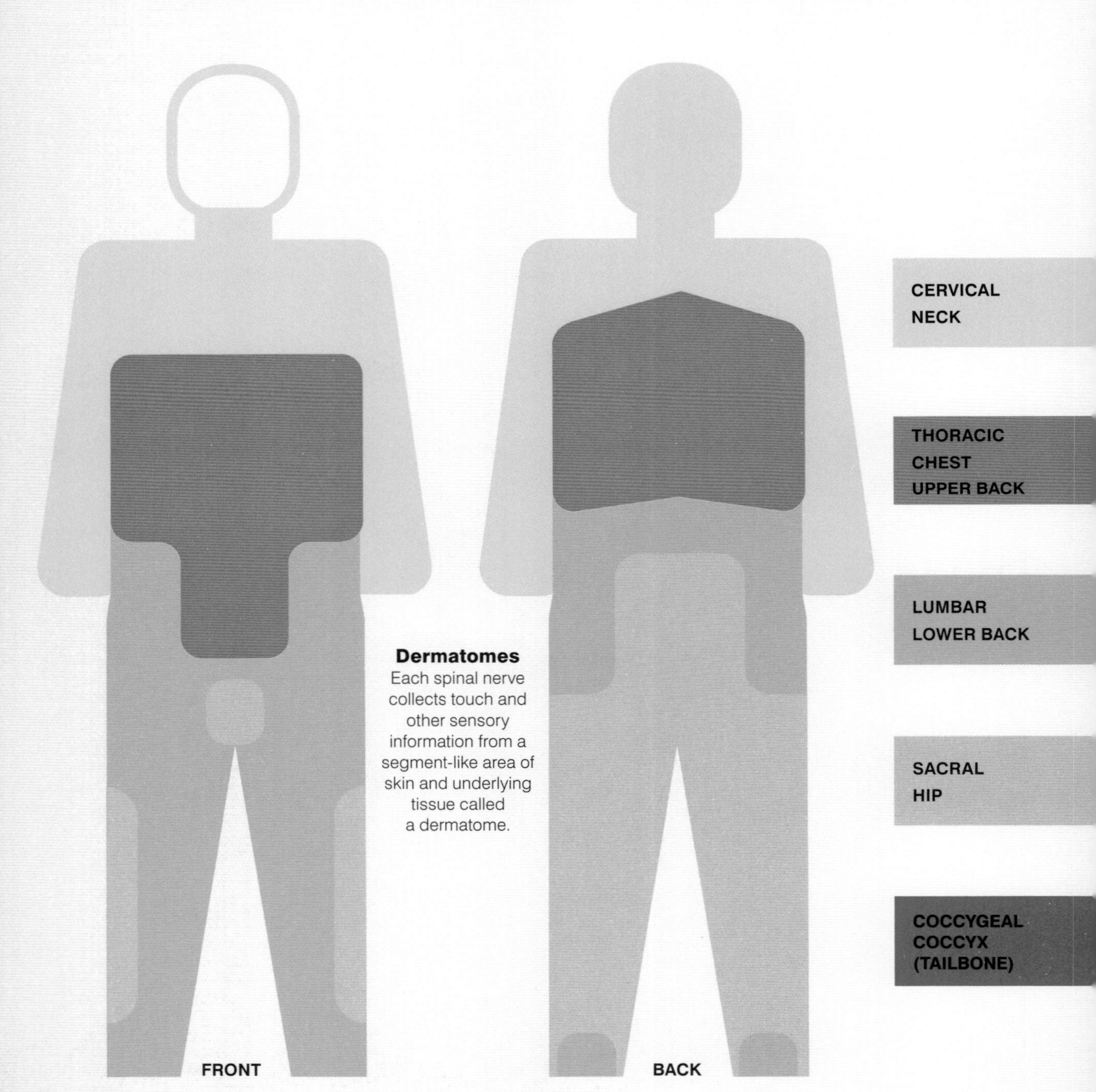

Dermatomes
Each spinal nerve collects touch and other sensory information from a segment-like area of skin and underlying tissue called a dermatome.

Spinal nerves
These nerves are named after their adjacent backbones:
1
2
3
4
5
6
7
8
1
2
3
4
5
6
7
8
9
10
11
12
1
2
3
4
5
1
2
3
4
5
1

REFLEXES AND REACTIONS

Often the brain's attention must focus on a particularly important task, like reading this book – or flying a supersonic jet plane. So as not to interrupt, many body parts look after themselves with automatic movements called reflexes. The part responds to a stimulus, such as a touch, by nerve signals that go to the spinal cord and 'short-circuit' straight back to muscles to make the necessary movement, leaving the brain to catch up later if needed. Reactions are quick, purposeful movements that do involve the brain's conscious alertness as it detects a situation, thinks fast and orders a rapid response.

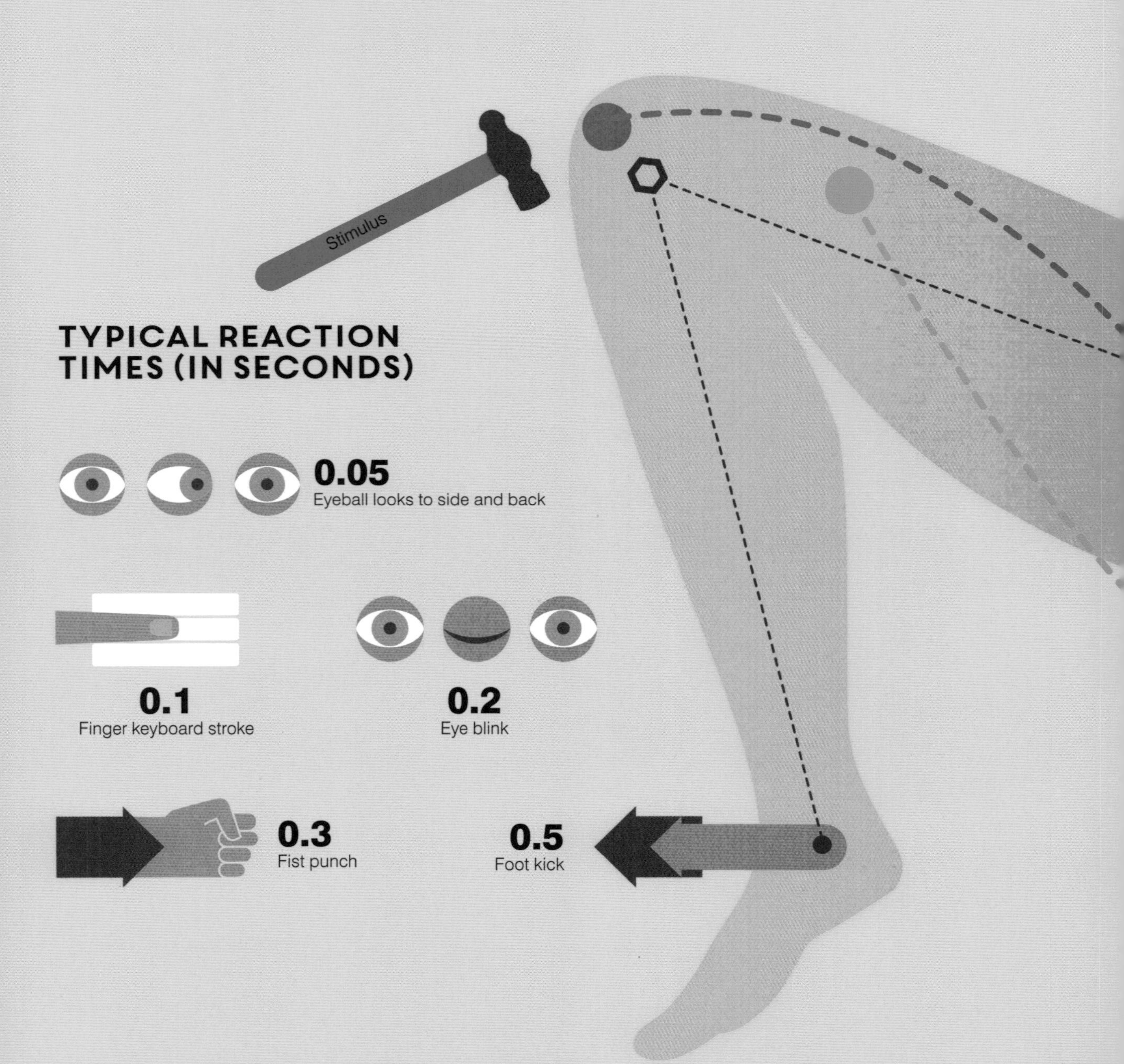

HOW REFLEXES HAPPEN

The body senses a stimulus such as sudden movement, an unfamiliar touch or pain, and orders an action straight away. Nerve signals also go to the brain, where they are filtered subconsciously to see if they are important enough to enter conscious awareness.

Sensory nerve

Relay nerve

Motor nerve

Relay up spinal cord

POINT TO THE ODD-ONE-OUT

Pointing to the odd-one-out of three objects **0.7** seconds

Pointing to the odd-one-out of six objects **1** second

PANS: NORMAL SERVICE

The parasympathetic ANS carries out everyday 'housekeeping'. Most of its nerve control comes from the brain via the spinal cord. Its actions usually oppose the sympathetic part. In daily life the two continually balance their effects.

SANS: EMERGENCY!

The sympathetic part of the ANS prepares the body for action, energy use and self-protection, often called 'fright, fight or flight', along with the hormone system. Most of its control is by the vagus nerve and sympathetic nerve chains (ganglia) alongside the spinal cord.

CRANIAL

CERVICAL

THORACAL

LUMBAL

BLOOD GLUCOSE (SUGAR)
Usual amounts available for energy

EYE PUPILS
Constricted (narrowed)

DIGESTIVE ACTIVITY
Appropriate

BLOOD PRESSURE
Standard

HEARTBEAT
Normal

BREATHING
Steady

MUSCLES
Relaxed

BLOOD GLUCOSE (SUGAR)
More available for energy

EYE PUPILS
Dilated (widened)

DIGESTIVE ACTIVITY
Reduced

BLOOD PRESSURE
Raised

HEARTBEAT
More rapid

BREATHING
Faster, deeper

MUSCLES
Tensed, ready with extra blood supply

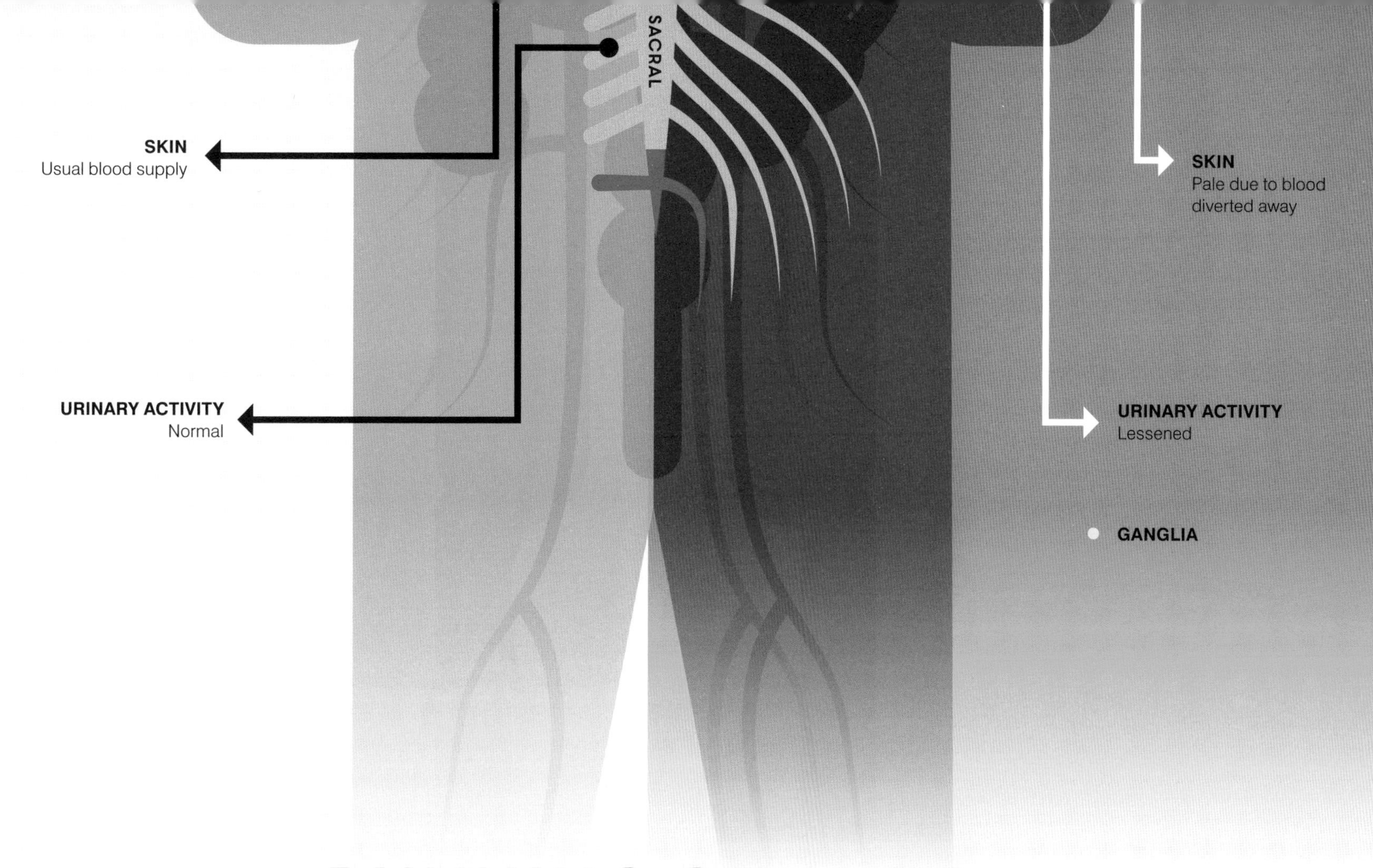

RUNNING ON AUTO

The human brain is truly astonishing, but even it has limited processing power for information passing through at the level of conscious awareness. Therefore it offloads much of the body's inner workings, like digesting food, heartbeat, breathing and waste collection, to happen automatically under the auspices of the autonomic nervous system or ANS. This is part of the peripheral system and organizes internal goings-on by itself, subconsciously – only alerting the thinking, sentient level if something goes awry.

THE MASTER SWITCH

Working in concert with the brain and nerves is the second bodywide coordination-control-command system: hormonal or endocrine. It is based on natural chemicals called hormones made in parts of the body known as endocrine glands. The two systems are integrated by a grape-sized area of the lower front brain, the hypothalamus, and the baked-bean-like pituitary gland dangling below it. So the executive director and chief operating officer are a tasty duo.

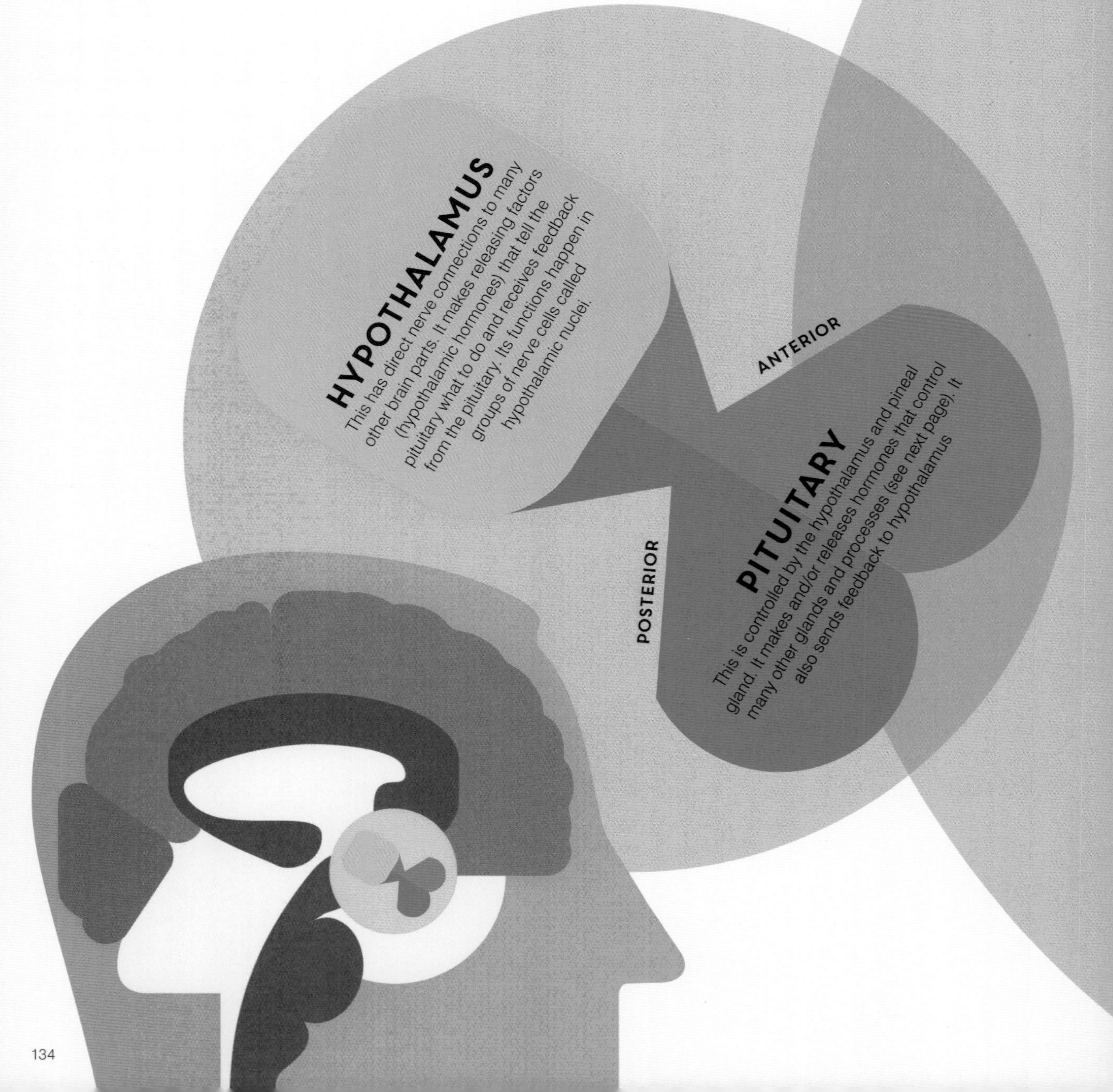

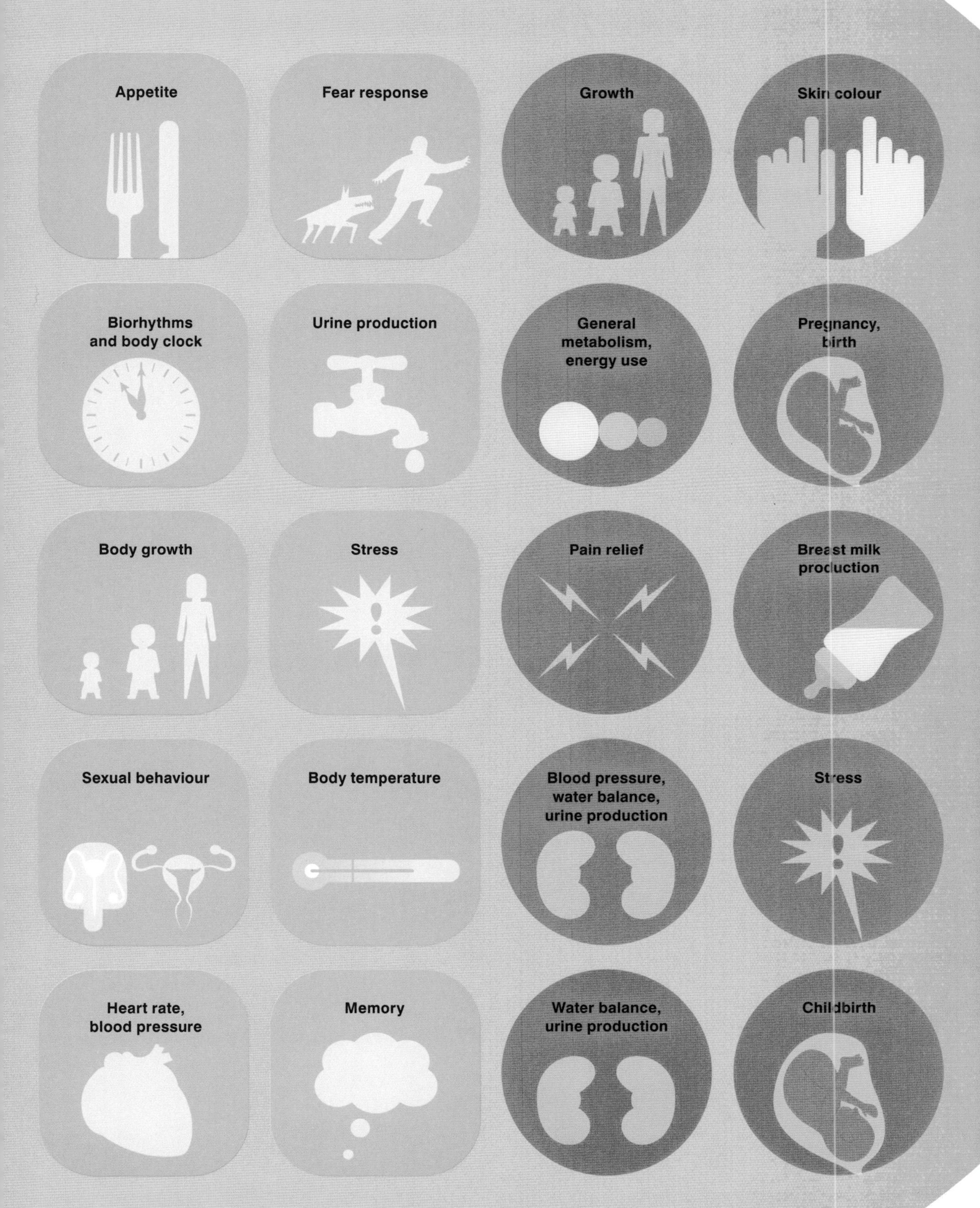
Appetite
Fear response
Growth
Skin colour
Biorhythms and body clock
Urine production
General metabolism, energy use
Pregnancy, birth
Body growth
Stress
Pain relief
Breast milk production
Sexual behaviour
Body temperature
Blood pressure, water balance, urine production
Stress
Heart rate, blood pressure
Memory
Water balance, urine production
Childbirth

CHEMICALS IN CONTROL

Blood is not only nourishing and a distributor. It is also the massive highway network for hormones to spread around the body. Each hormone is a small, blood-borne chemical substance. It comes from a particular endocrine gland, reaches every nook and cranny, but only affects certain tissues and organs, known as its targets.

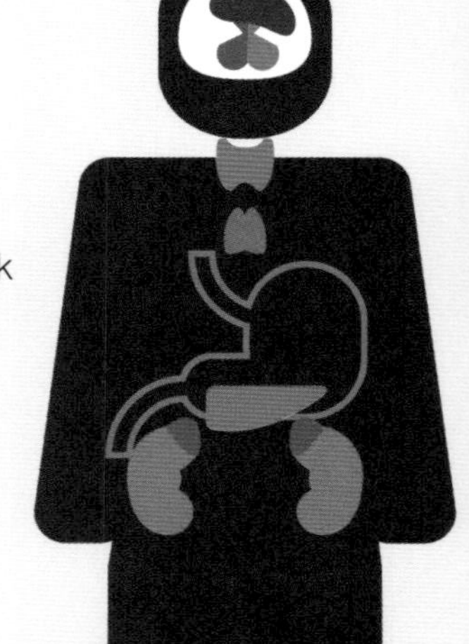

PITUITARY
'Master gland' of the hormone system

Products
Ten-plus hormones and similar substances (see previous page)

Targets
Most parts, from cells to large organs

Size
15 x 10 mm

PINEAL
Regulates sleep-wake patterns, biorhythms

Products
Melatonin

Targets
Most parts, especially brain

Size
9 x 6 mm

THYROID
Regulates metabolism, speed of body processes; controls blood calcium levels

Products
Thyroxine, triiodothyronine; calcitonin

Targets
Most cells of the body

Size
100 x 30 mm

PARATHYROIDS
Control blood calcium levels

Products
Parathyroid hormone

Targets
Most cells of the body

Size
6 x 4 mm

PANCREAS

Regulation of blood glucose (see next page)

Products

Insulin; glucagon

Targets

Most body cells

Size

13 x 4 cm

STOMACH

Release of acid and other digestive juices

Products

Gastrin; cholecystokinin; secretin

Targets

Stomach; pancreas, gall bladder; pancreas

Size

30 x 15 cm

ADRENAL 1: OUTER (CORTEX)

Regulates water and mineral levels; responses to stress; sexual development, activity

Products

Aldosterone; cortisol; sex hormones

Targets

Kidney and gut; most body parts; sex organs

Size

Whole gland 5 x 3 cm

ADRENAL 2: INNER (MEDULLA)

Prepares body for action (fright, fight, flight)

Products

Adrenaline and similar hormones

Targets

Many body parts

Size

Whole gland 5 x 3 cm

THYMUS

Stimulates white cells to fight disease

Products

Thymosin and similar hormones

Targets

White blood cells

Size

5 x 5 cm in childhood, shrinks in adult

KIDNEY

Water and mineral balance, blood pressure; red blood cell production

Products

Renin (an enzyme); erythropoietin

Targets

Kidney and blood circulation; bone marrow

Size

12 x 6 cm

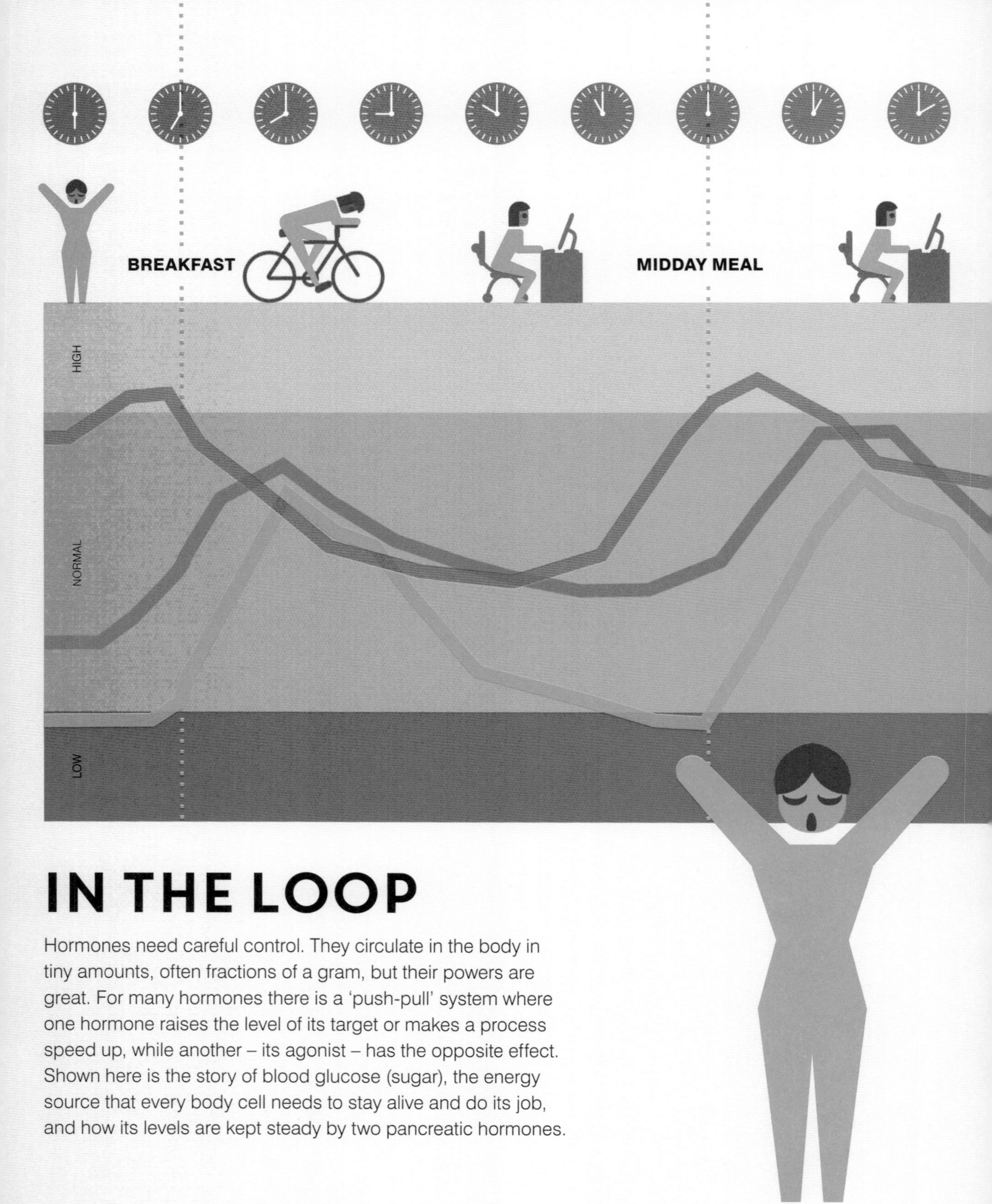

IN THE LOOP

Hormones need careful control. They circulate in the body in tiny amounts, often fractions of a gram, but their powers are great. For many hormones there is a 'push-pull' system where one hormone raises the level of its target or makes a process speed up, while another – its agonist – has the opposite effect. Shown here is the story of blood glucose (sugar), the energy source that every body cell needs to stay alive and do its job, and how its levels are kept steady by two pancreatic hormones.

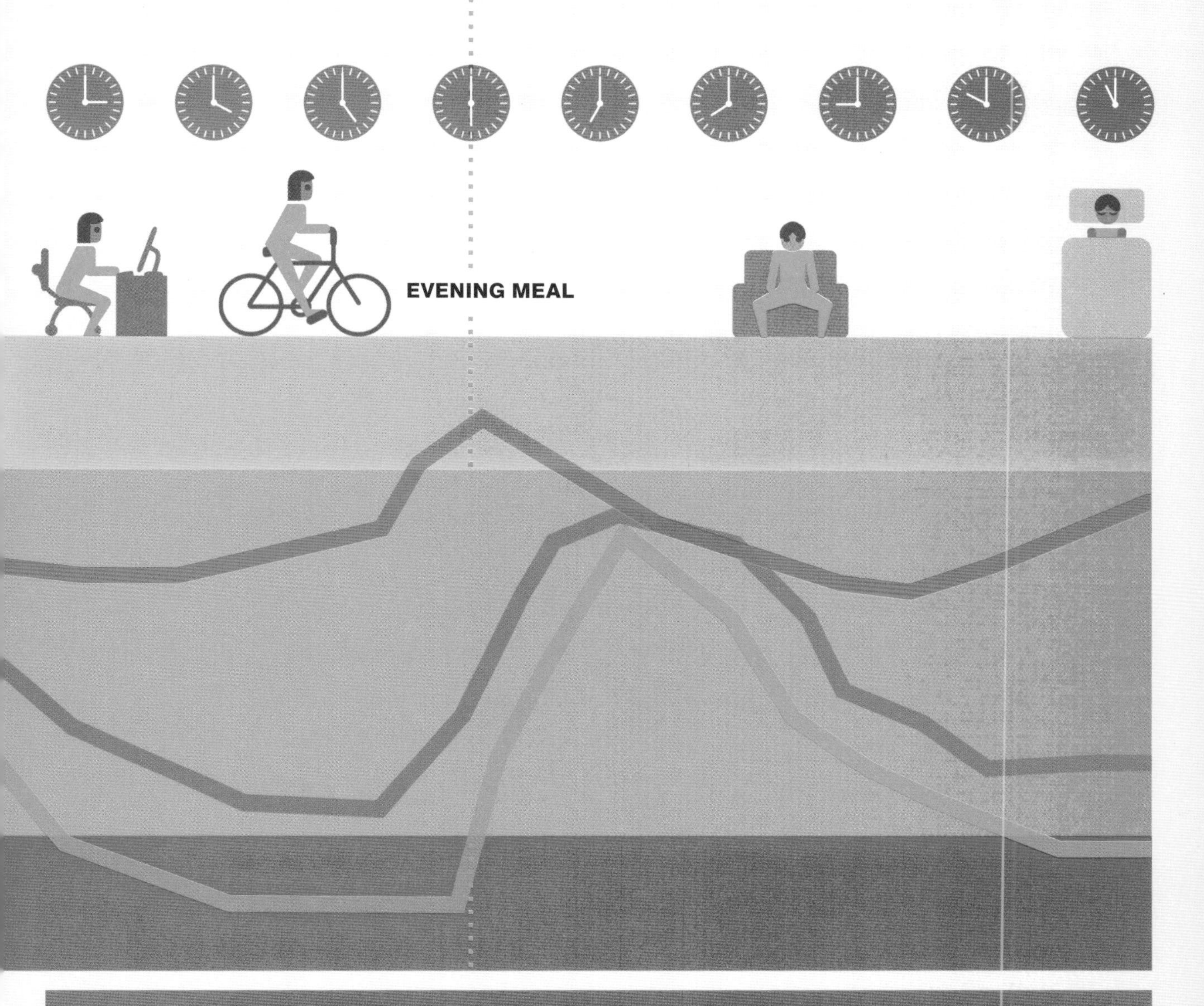

GLUCAGON
Source: Alpha cells in the islets of the pancreas.
Function: Raises level of glucose in blood by telling the liver to convert glycogen (starch) into glucose.
Levels: Glucagon goes down when the others go up, and after more of a time lag, 1–2 hours.

BLOOD GLUCOSE
Source: Food and drinks, especially sugary and starchy ones (carbohydrates).
Function: Provides energy for metabolic processes in every cell.
Levels: Rise after food is eaten (especially with high carb foods) and fall with activity and exercise.

INSULIN
Source: Beta cells in the islets of the pancreas.
Function: Lowers level of glucose in blood by encouraging its take-up by cells and its conversion into glycogen in the liver.
Levels: Insulin follows glucose, a few minutes after.

GOING STEADY

A balance of water and minerals is vital to good health. As the body eats, drinks, breathes, sweats, exercises and does almost anything else, the balance is liable to become disturbed. Several parts and hormones work together to make sure this does not happen and preserve the status quo.

HYPOTHALAMUS
Senses levels of water and minerals in blood, makes some hormones, including ADH (anti-diuretic hormone, also called vasopressin)

PITUITARY
Makes, stores, releases hormones, including ADH

Kidneys
Produce renin. Filter wastes from blood. Contain about a million micro-filters called nephrons

Wastes, water and minerals filtered out into tubule

Under hormone control (ADH, aldosterone, ANP) some water and minerals are taken back into blood, according to body's needs

Unfiltered blood flows through knot of capillaries

Urine out to bladder

LOW BLOOD PRESSURE

When levels of water in blood decrease and blood pressure falls

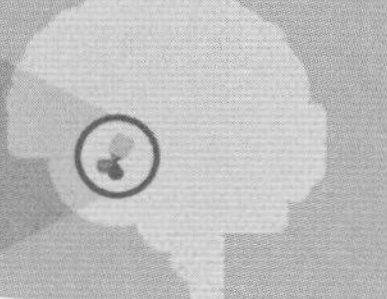

Pituitary gland releases ADH (antidiuretic hormone or vasopressin)

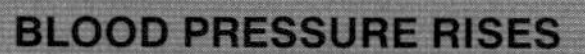

BLOOD PRESSURE RISES

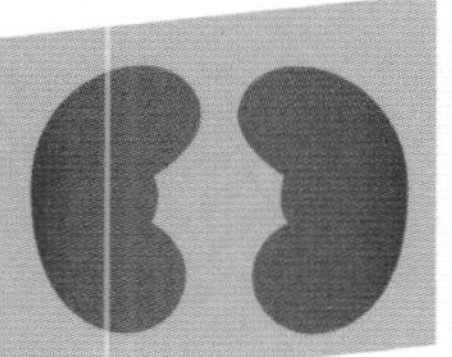

Renin released by kidneys converts AT1 (angiotensin 1) from liver to AT2

ADH targets kidneys to make them remove more water from urine into the blood.

Narrower blood vessels and more water in blood

AT2 makes blood vessels narrower to raise blood pressure, and stimulates adrenal gland to release aldosterone.

ADH also makes blood vessels narrower to raise blood pressure

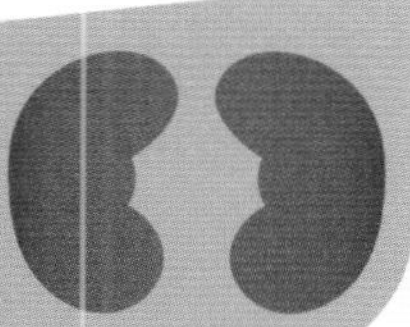

Aldosterone targets kidneys so they remove more water from urine into blood

Targets kidneys to make them remove less water from urine

Less water from kidneys goes into blood so blood volume decreases

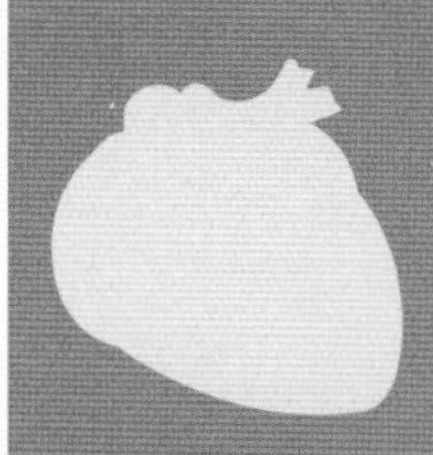

ANP (atrial natriuretic peptide) made in atria (upper chambers) of heart is released

BLOOD PRESSURE LOWERS

HIGH BLOOD PRESSURE

When levels of water in blood increase and blood vessels are narrowed

THINKING BODY

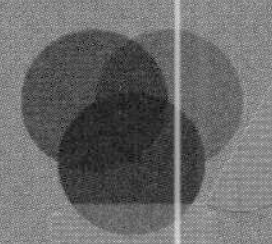

BRAIN BY NUMBERS

There is a wide range of sizes for normal brains (broad averages are shown here), and no simple link between these overall dimensions and intelligence. Despite the brain's still, inert appearance, it buzzes with nervous electrical and chemical activity, making it – on average – the most energy-hungry organ in the entire body.

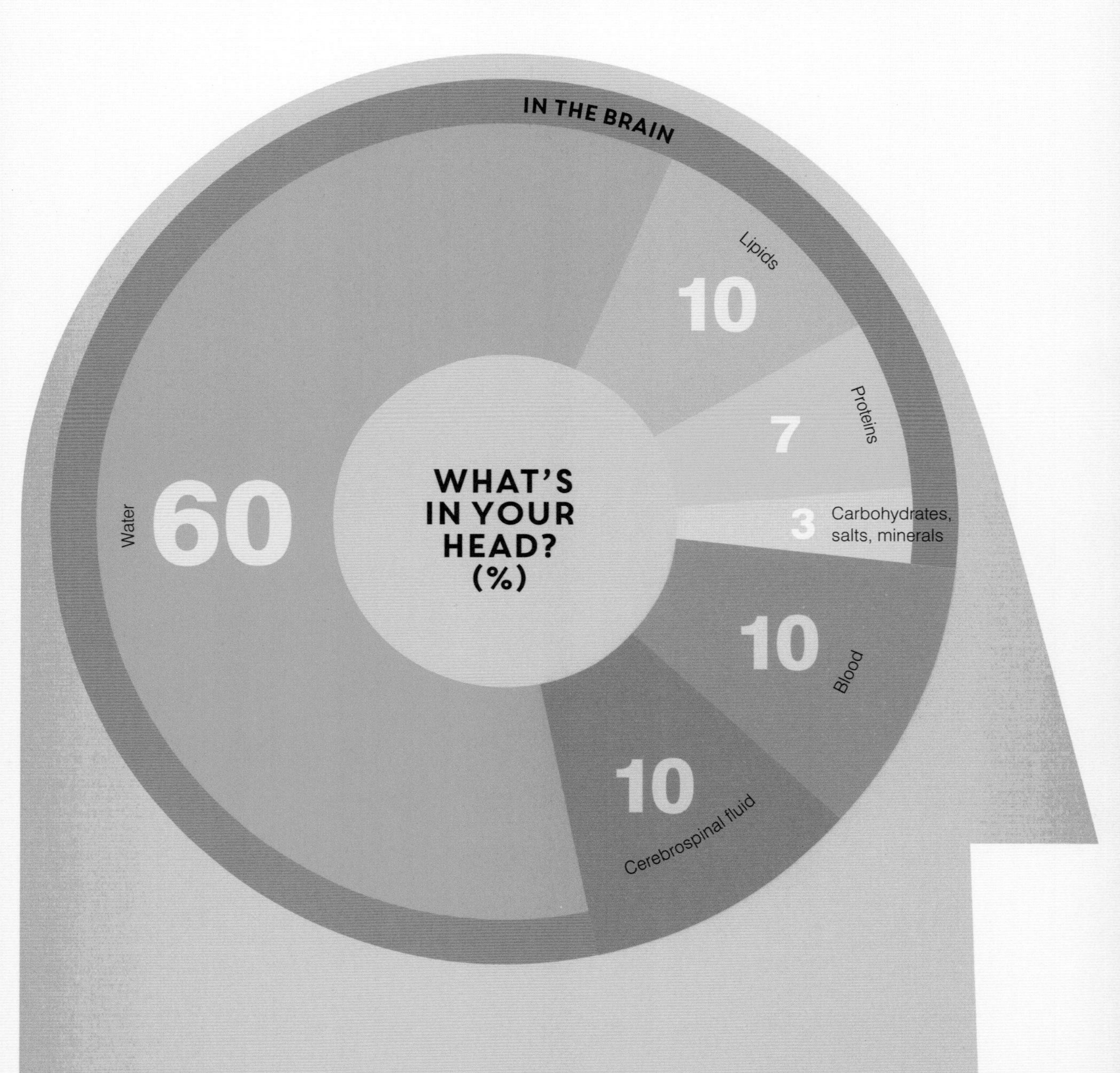

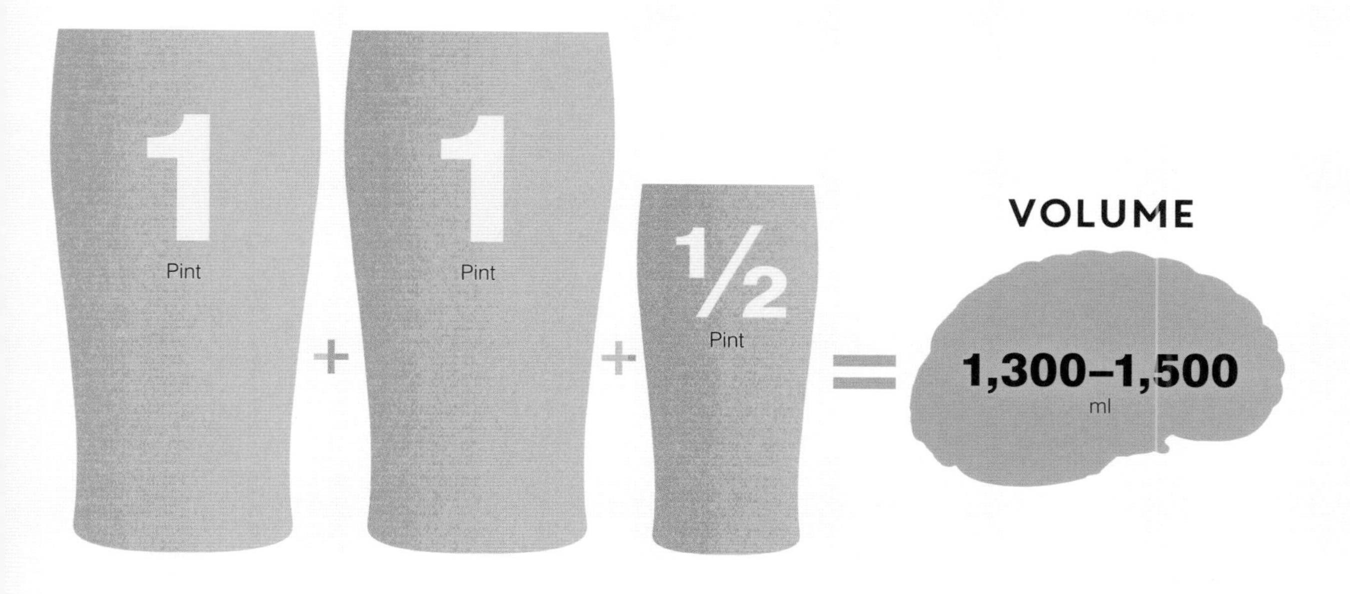
1
Pint
+
1
Pint
+
½
Pint
=
VOLUME
1,300–1,500
ml

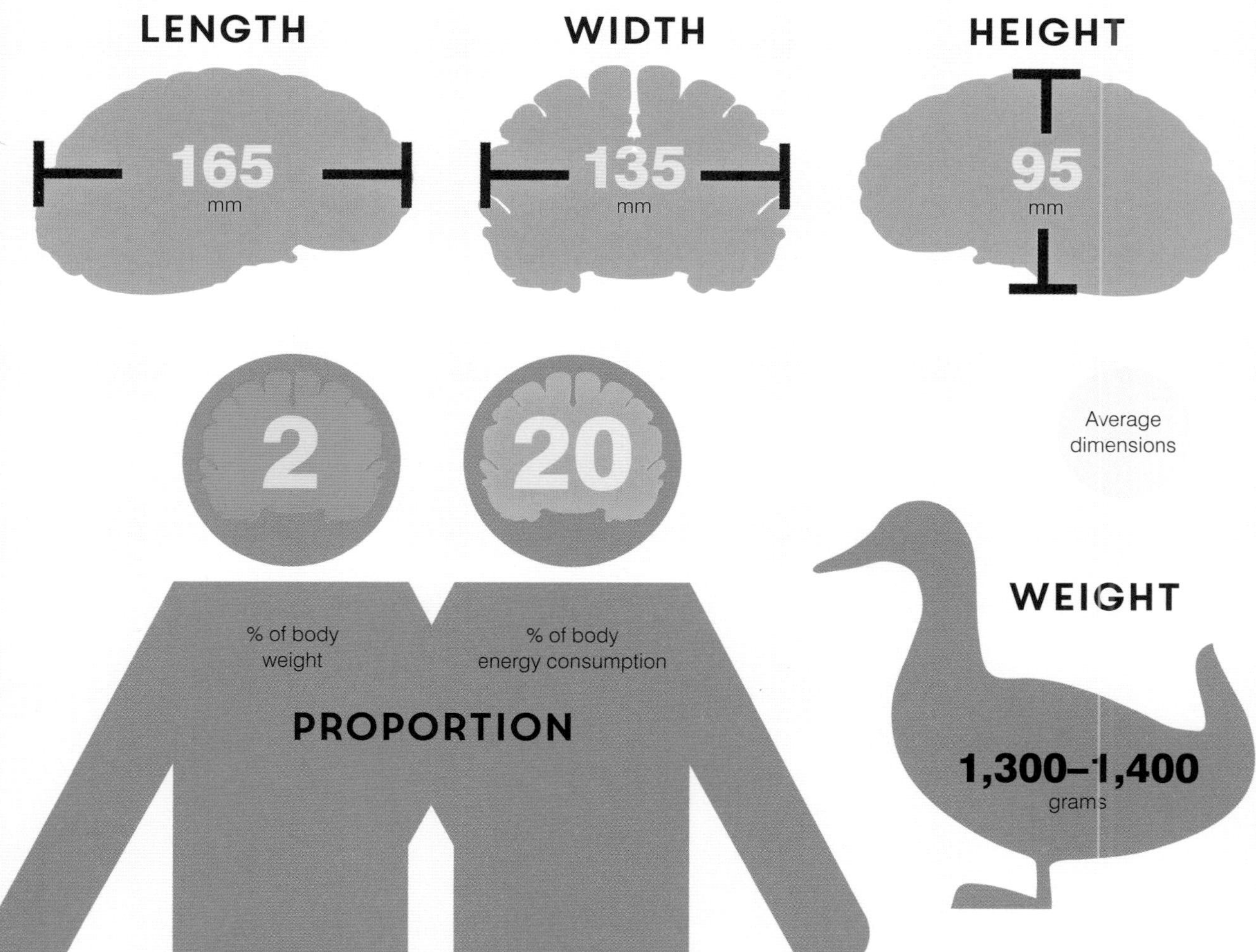
LENGTH
165
mm
WIDTH
135
mm
HEIGHT
95
mm
2
% of body weight
20
% of body energy consumption
PROPORTION
Average dimensions
WEIGHT
1,300–1,400
grams

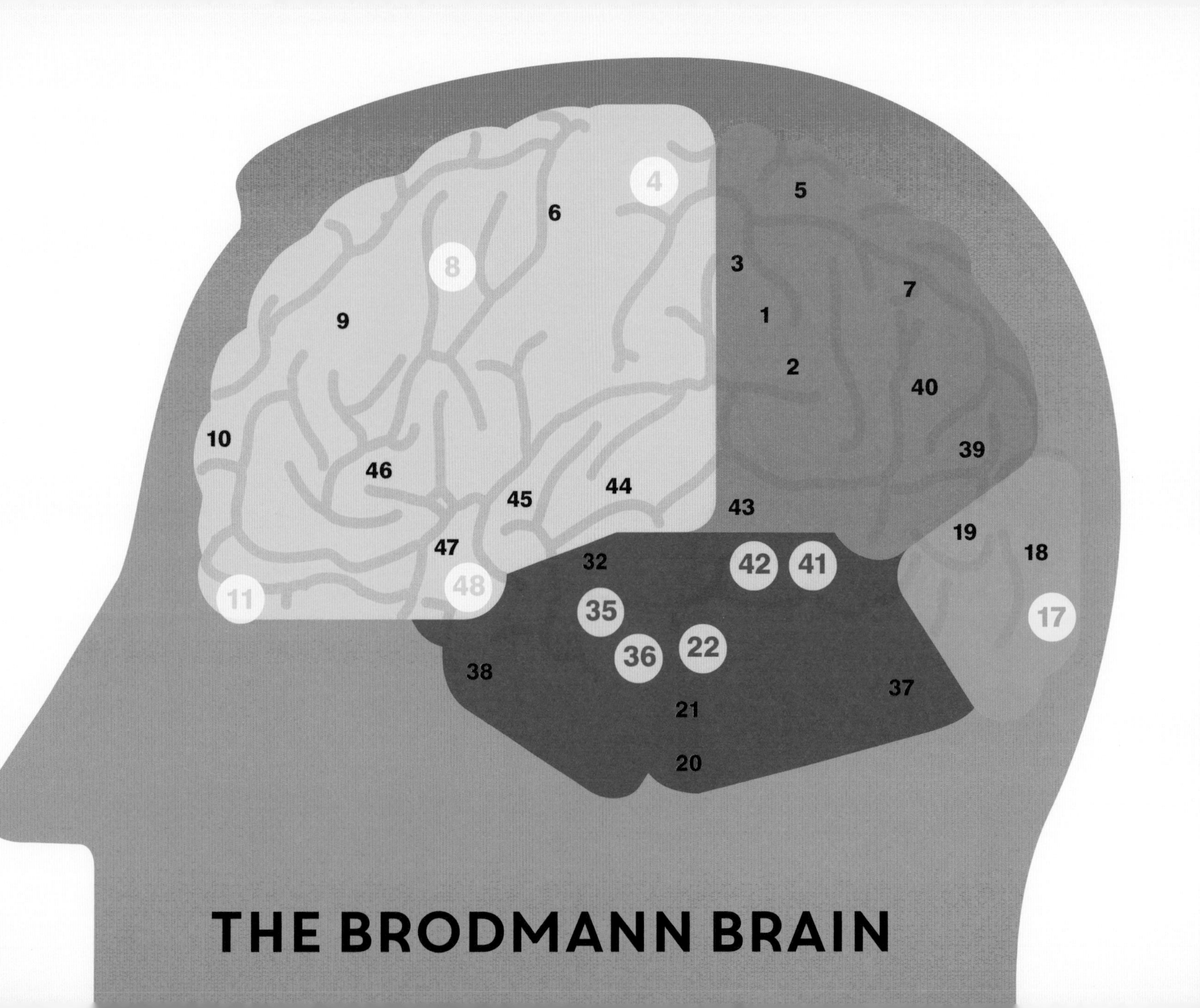

THE BRODMANN BRAIN
4
5
6
8
3
7
9
1
2
40
10
39
46
45
44
43
47
19
18
32
42
41
48
11
35
17
36
22
38
37
21
20

Look very, very closely at the cortex (the main folded surface of the brain). Its microscopic nerve cells are not all the same. Their shapes, numbers, sizes, and six-layered organization vary, like a patchwork. The patches are called Brodmann areas, and each has its own number and its own roles. A selection of the main areas and their function is shown here.

4 MOVEMENTS **Primary motor cortex**
Orders contractions of muscles to make movements.

8 DECISIONS **Prefrontal cortex**
Among several sites involved in doubts, decisions and uncertainty.

11 REWARD **Prefrontal cortex**
Among several sites for making decisions, assessing rewards, reasoning and long-term memory.

17 SIGHT **Primary visual cortex**
Main destination for messages from the eyes concerning vision.

22 SPEECH **Understanding words**
Wernicke's area (left side), ambiguity (right side).

35, 36 SIGHT & MEMORY **Temporal cortex**
Allows seen objects to be recognized and given meaning.

41, 42 HEARING **Primary auditory cortex**
Main destination for messages from the ears concerning sounds.

48 AWARENESS **Prefrontal cortex**
Among several sites for working memory, awareness, attention and concentration.

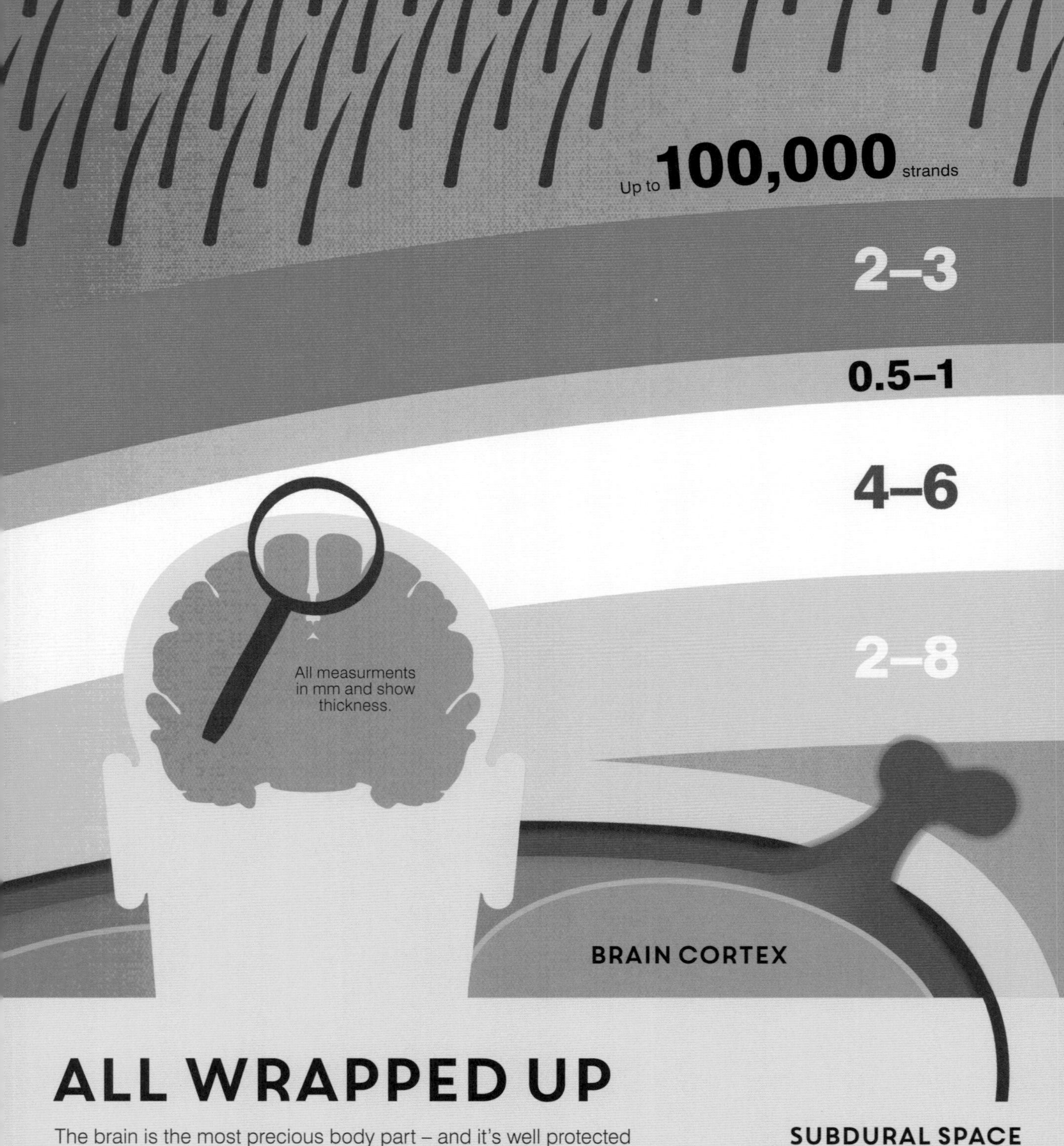

ALL WRAPPED UP

The brain is the most precious body part – and it's well protected by lots of natural layers around it. These provide a delicate, intermeshed combination of strength, security, cushioning and flexibility. The three main layers are the dura, arachnoid and pia, collectively called the meninges. (One layer, singular, is a meninx.) Extras can be added on the outside, like a hard hat …

SUBDURAL SPACE

This is 'potential space' since the dura is usually attached to the arachnoid, and the dura and arachnoid separate only with problems (disease, injury).

HAIR

These are keratin-based strands. Each strand self-renews after 3–5 years.

SCALP SKIN

Chiefly made up of collagen, elastin and keratin proteins, it is self-renewing after 4 weeks.

PERIOSTEUM

Tough outer 'skin' covering bone tissue.

SKULL BONE

The cranium of the skull, the part covering the brain, consists of eight cranial bones connected by firm, fused joints called sutures.

MENINGES 1: DURA MATER

Translated as 'Tough Mother', this layer is a tough, strong outer casing for the other meninges and brain. It consists of dense fibres arranged in layers called laminae, and supports blood vessels and varied blood spaces (sinuses).

0.1–3

MENINGES 2: ARACHNOID MATER

The 'Spider Mother' layer is a delicate spongy web of collagen and other connective tissue, and fluids. It is a flexible foam-like cushion that absorbs the impact of shocks to the head

0.1

MENINGES 3: PIA MATER

The mesh-like fibre network of the 'Tender Mother' is the last line of protection against contact with the cortex, and closely follows the contours of the brain surface.

0.3–8

SUB-ARACHNOID SPACE

This contains cerebrospinal fluid and is a flowable fluid cushion that absorbs impact shocks to head.

CUTAWAY BRAIN

It's not much to look at – a wrinkled lump of grey and white with a few curved squiggly bits inside. Yet this is the control centre for the physical body, the chief coordinator for the chemical body, the mental body's seat of the mind, the storehouse for memories, the origin of emotions, and the second-by-second hub for conscious awareness.

CEREBRUM

Large upper wrinkled dome divided into two hemispheres. Forms 80% of whole brain by volume.
Contents: Mostly white matter, nerve fibres (axons)
Function: Links cortex to rest of brain

CORPUS CALLOSUM

10 cm long connecting strap between left and right hemispheres
Contents: More than 200 million nerve fibres
Function: Lets each side of the body know what the other side is doing – literally

MIDBRAIN

Forms 10% of whole brain by volume
Contents: Mix of nerve cells and fibres
Function: Mainly concerned with automatic body maintenance

THALAMUS

Twinned egg-shaped lumps, 5–6 cm long
Contents: Nerve cells and fibres organized in areas called nuclei
Function: 'Gatekeeper' to the cortex and conscious mind

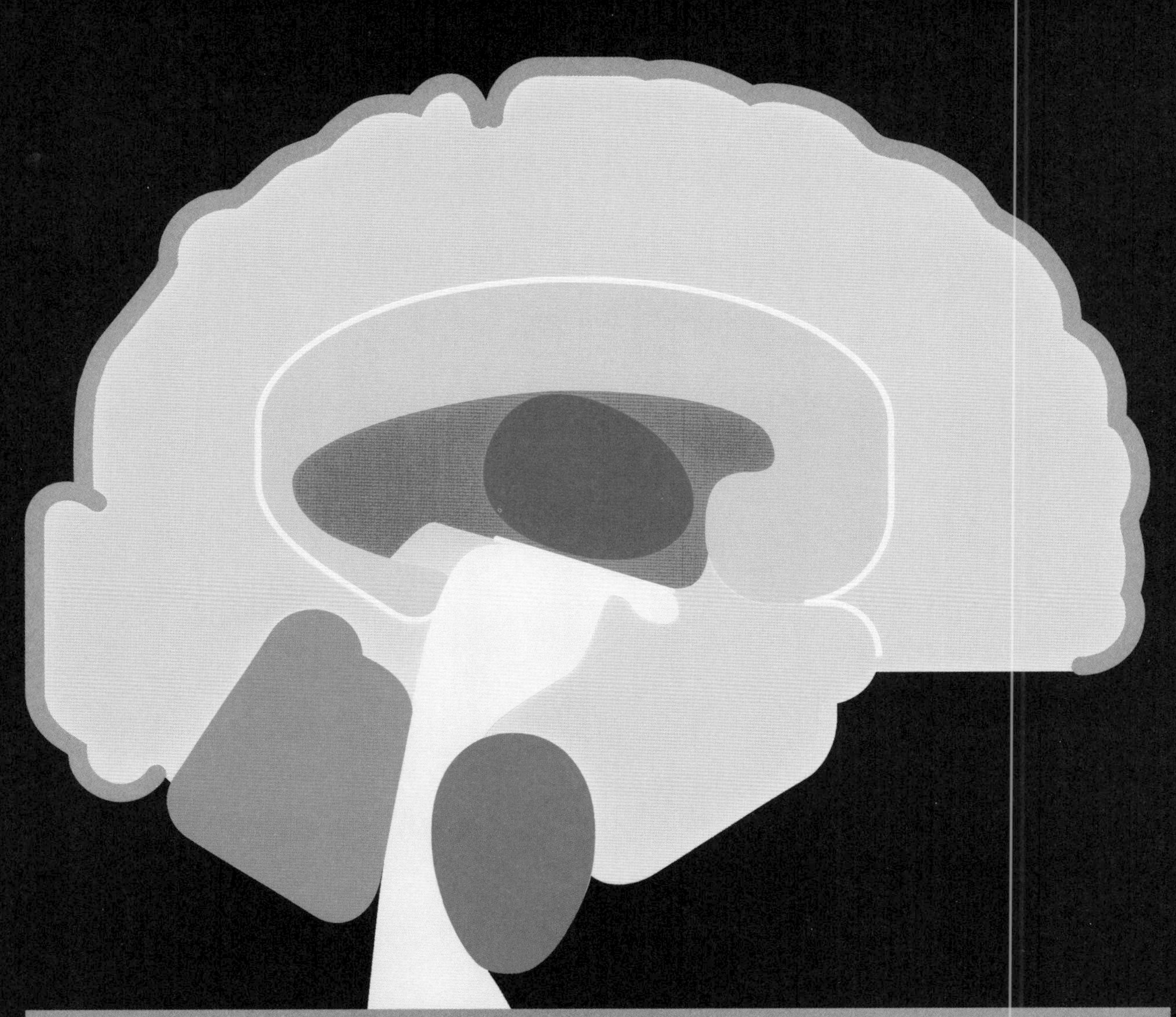

CEREBRAL CORTEX	Thin grey covering of cerebrum **Contents:** 20 billion nerve cells (neurons) **Function:** Site of awareness and most conscious thought processes
BRAIN STEM	Lowest part of the brain, merging into spinal cord below **Contents:** Mix of nerve cells and fibres **Function:** Site of centres for basic body processes such as breathing, heartbeat (see pages 128/132)
PONS	Size top-bottom 2–3 cm **Contents:** Mainly nerve fibres **Function:** Link between lower and higher brain parts
CEREBELLUM OR 'LITTLE BRAIN'	Forms 10% of whole brain by volume **Contents:** 50-plus billion nerve fibres **Function:** Involved in movement and coordination (see next page)

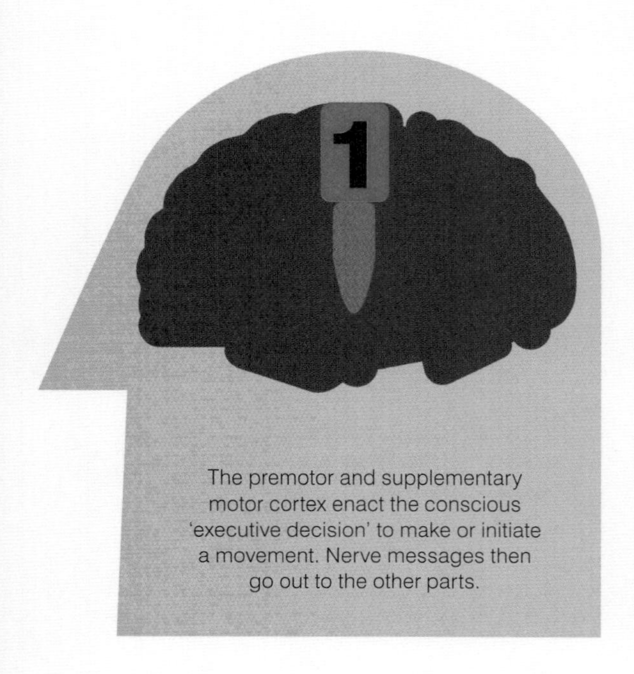

The premotor and supplementary motor cortex enact the conscious 'executive decision' to make or initiate a movement. Nerve messages then go out to the other parts.

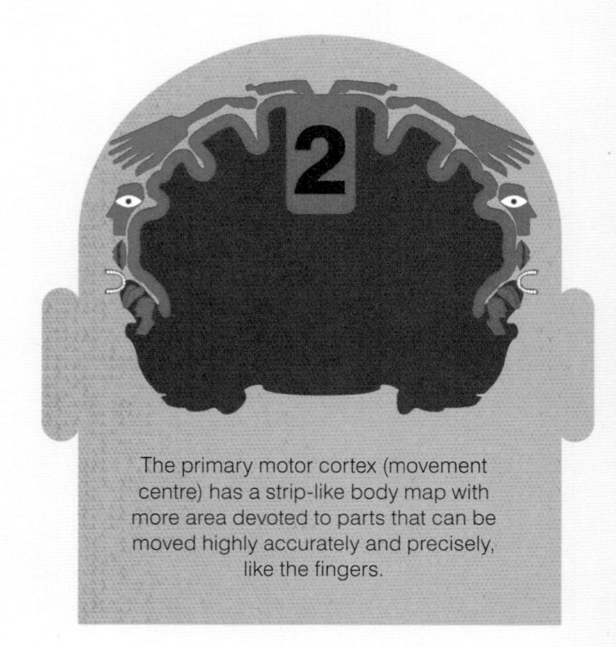

The primary motor cortex (movement centre) has a strip-like body map with more area devoted to parts that can be moved highly accurately and precisely, like the fingers.

MAKE YOUR MOVE

Movements seem so simple. The brain thinks about them – and they happen. But the process involves several brain parts sending messages to and fro between them, especially strips on the surface called the motor cortexes, also the cerebellum at the rear, thalamus in the centre, small parts deep in the brain called basal ganglia, and others. Then there's the journey from the brain along nerves to the muscles, to make them contract, which pulls the bone and makes them move. So overall, not quite that simple...

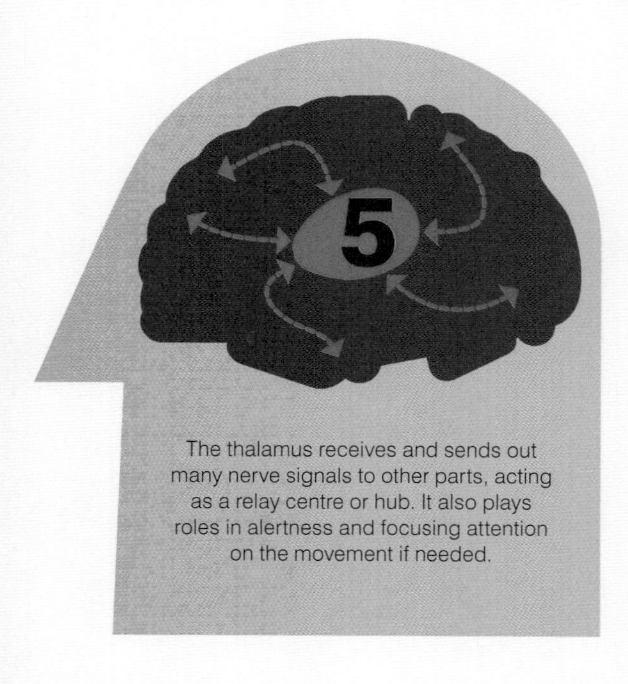

The thalamus receives and sends out many nerve signals to other parts, acting as a relay centre or hub. It also plays roles in alertness and focusing attention on the movement if needed.

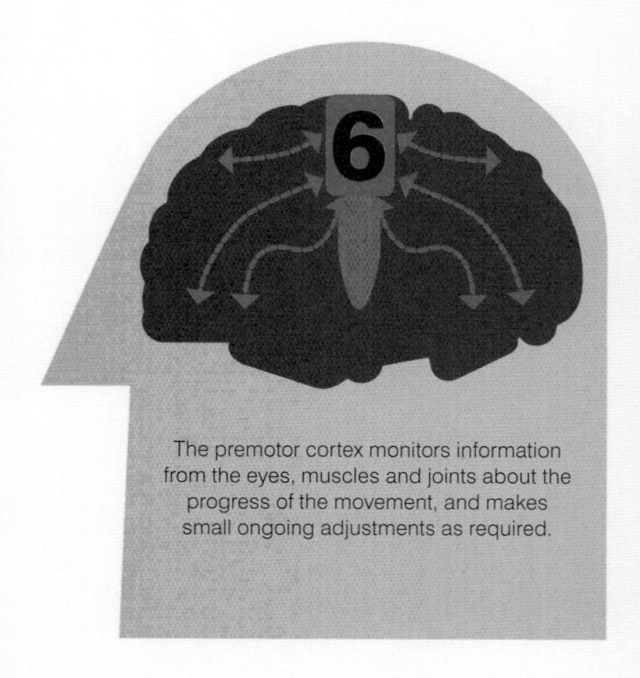

The premotor cortex monitors information from the eyes, muscles and joints about the progress of the movement, and makes small ongoing adjustments as required.

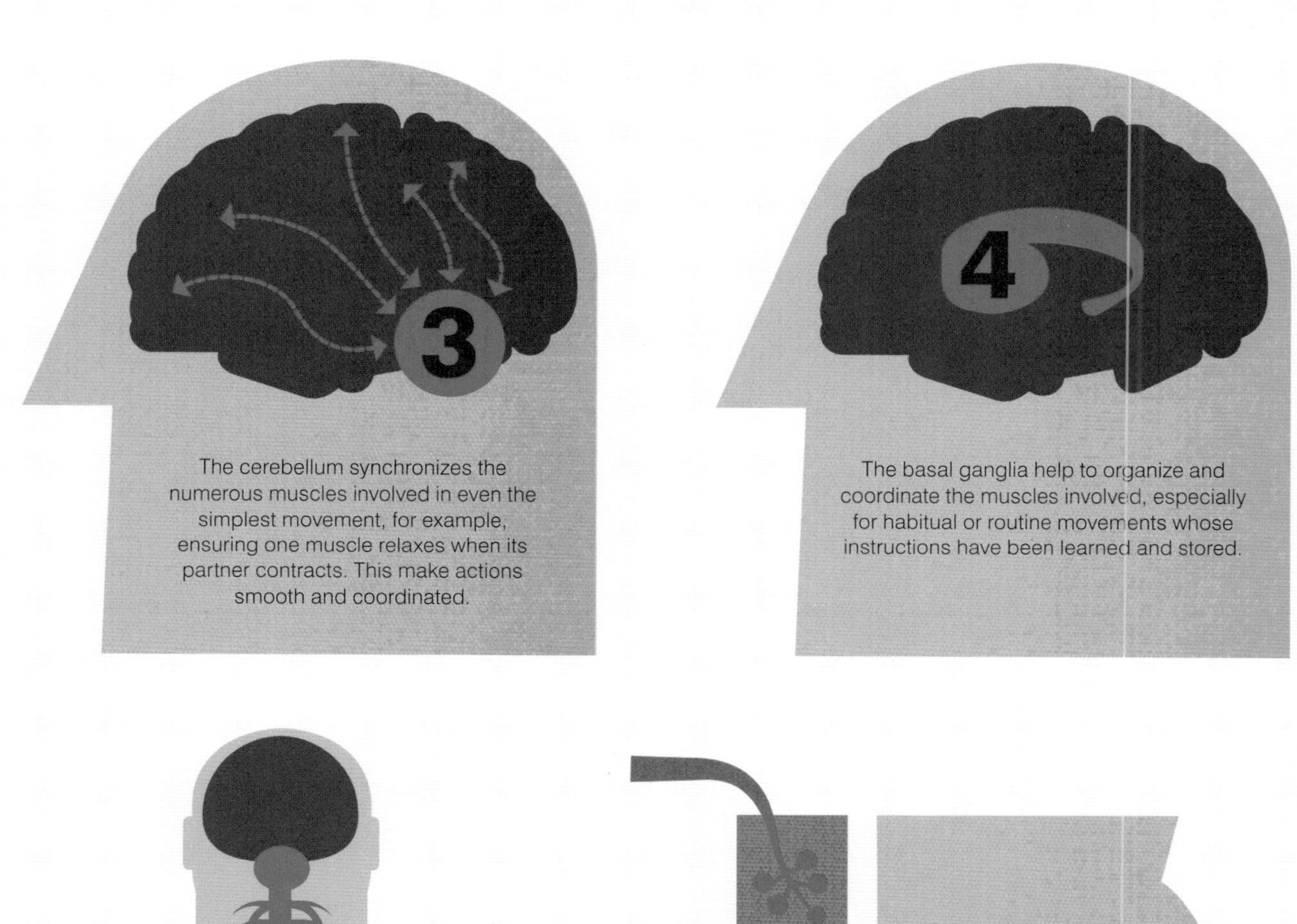
3
The cerebellum synchronizes the numerous muscles involved in even the simplest movement, for example, ensuring one muscle relaxes when its partner contracts. This make actions smooth and coordinated.
4
The basal ganglia help to organize and coordinate the muscles involved, especially for habitual or routine movements whose instructions have been learned and stored.

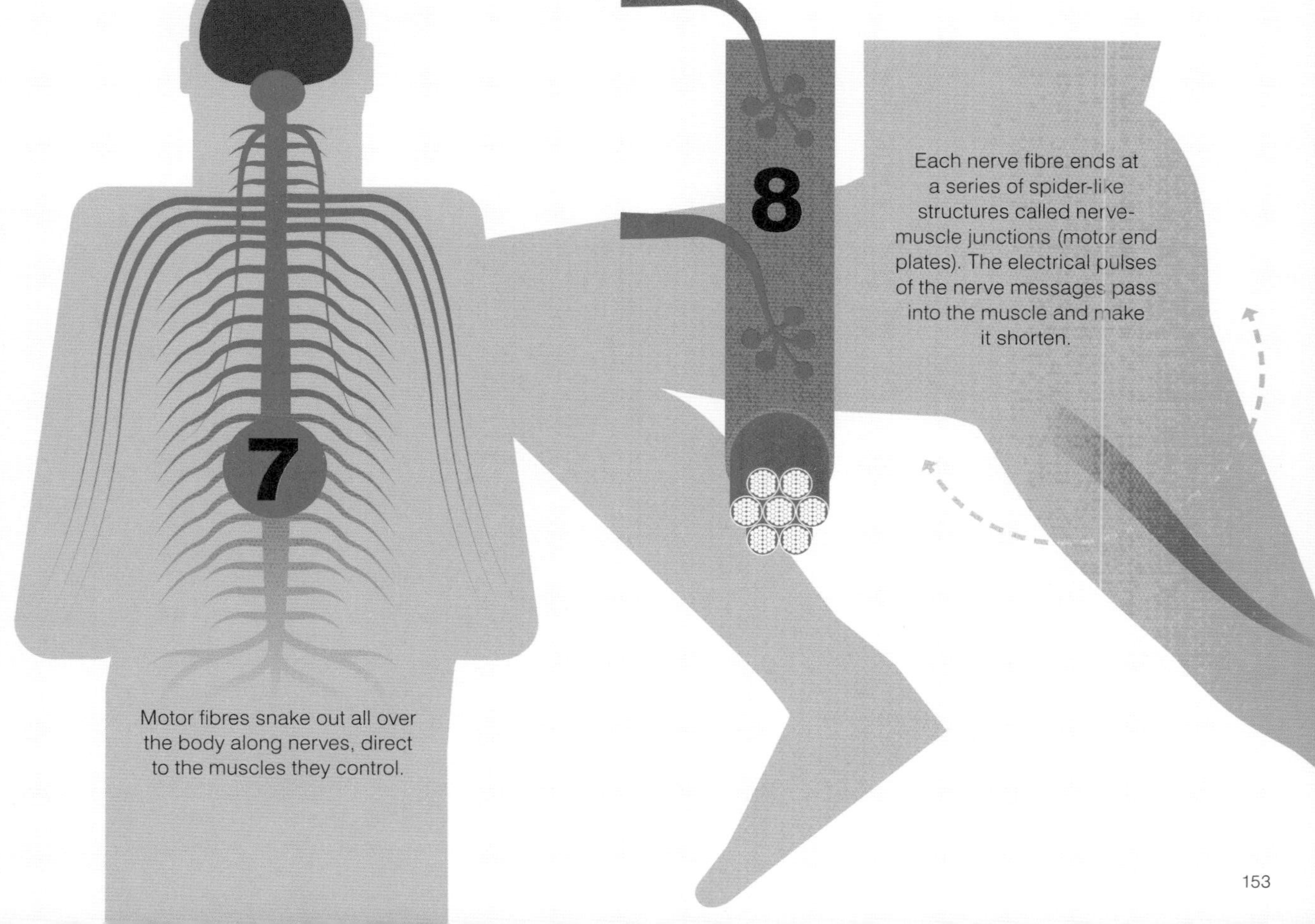
7
Motor fibres snake out all over the body along nerves, direct to the muscles they control.
8
Each nerve fibre ends at a series of spider-like structures called nerve-muscle junctions (motor end plates). The electrical pulses of the nerve messages pass into the muscle and make it shorten.

LEFT OR RIGHT?

The two sides of the brain look almost identical. But in the way they work and what they control, they are not the same. Some of these differences relate to whether a person is right- or left-handed. Some are linked to how the brain learns to do various tasks. And some differences are 'hard-wired' into the brain's nerve circuits. The general term for this is lateralization of brain function – although more and more research shows the differences are more complex than once thought.

13 August each year is Left-hander's Day

Selfish
Tends to interact with itself more than with the right hemisphere.

Talkative
Tends to dominate in language, vocabulary, syntax, grammar, especially in right-handed people.

In most human communities, an average of **1 in 10 people is left-handed,** which means they preferentially use their left hand, especially for dextrous tasks and delicate manipulation. However, this average disguises is a wide range, from 1 in 4 to 1 in 50.

Despite many myths to the contrary, there's little hard evidence to show that a higher proportion of artistic, musical and generally creative people are left-handed.

Left-handers tend to perform manipulative tasks better with their right hands than right-handers do with their left hands.

Hard
Often said to take charge in analytical 'hard' processes such as numerical tasks, calculations, formulae, logic, step-by-step reasoning, categorizing, definitions, efficiency, science and technology.

BUT recent research shows this is less

1 2 3 4 5 6 7 8 9 10

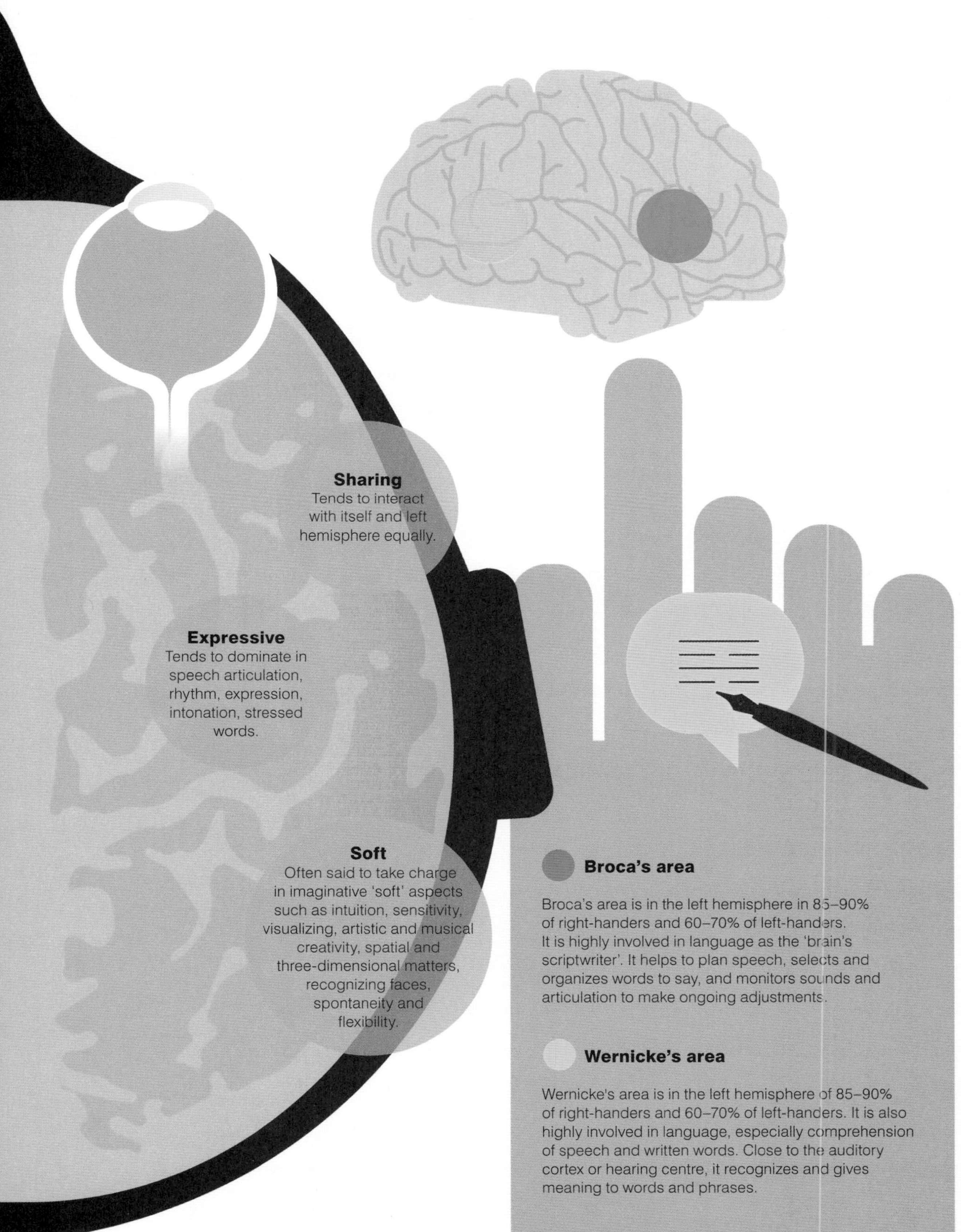
Sharing
Tends to interact with itself and left hemisphere equally.
Expressive
Tends to dominate in speech articulation, rhythm, expression, intonation, stressed words.
Soft
Often said to take charge in imaginative 'soft' aspects such as intuition, sensitivity, visualizing, artistic and musical creativity, spatial and three-dimensional matters, recognizing faces, spontaneity and flexibility.
Broca's area
Broca's area is in the left hemisphere in 85–90% of right-handers and 60–70% of left-handers. It is highly involved in language as the 'brain's scriptwriter'. It helps to plan speech, selects and organizes words to say, and monitors sounds and articulation to make ongoing adjustments.
Wernicke's area
Wernicke's area is in the left hemisphere of 85–90% of right-handers and 60–70% of left-handers. It is also highly involved in language, especially comprehension of speech and written words. Close to the auditory cortex or hearing centre, it recognizes and gives meaning to words and phrases.

THE FLUID BRAIN

It's official: the brain is mostly mush. About 75% of this most vital organ is water, chiefly found in and between cells. Aside from the brain, almost all the other skull contents are water-based too. The main liquids here are blood and a curious substance unique to the nervous system known as CSF, cerebrospinal fluid, which circulates slowly through chambers called ventricles – because the brain is hollow!

CSF AND THE BRAIN | BLOOD AND THE BRAIN

Volume in brain at any moment (ml)

CSF	Blood
150	120

CSF provides physical protection and cushioning; removes wastes; helps regulate blood pressure in the brain; and supplies some nutrients.

Origin: choroid plexuses lining brain ventricles.

Fate: Absorbed in subarachnoid space and by veins.

Blood delivers oxygen, energy (glucose), nutrients and minerals; removes wastes; distributes warmth; and fights infection.

Origin: heart's left ventricle via internal carotid (80%) and vertebral (20%) arteries.

Fate: heart's right ventricle via jugular veins.

Measured in ml

Brain ventricles	Brain's sub-arachnoid space	In and around spinal cord	Arteries	Capillaries and brain tissue	Veins
30	120	50	20	55	45

In the brain

1

Red
blood
cells

2

3

THE 'TRIPLE-B'

The brain has special protection from nasties in the blood like many kinds of germs and toxic chemicals. This is the blood-brain barrier, the triple-B. It's based on three differences between brain capillaries and ordinary capillaries around the rest of the body.

1 **Between cells forming capillary wall**
In the brain: No gap
Rest of body: Gaps

2 **Capillary wall base membrane**
In the brain: Continuous
Rest of body: Gaps

3 **Protective cells around capillary**
In the brain: Protective astrocyte cells
Rest of body: None

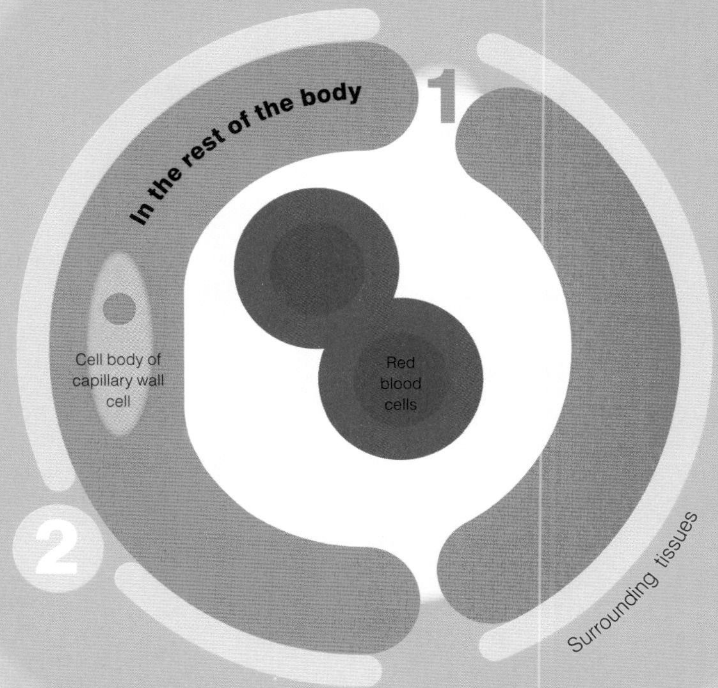

IN-HEAD INTERNET

The brain's main microfeatures are nerve cells, or neurons – more than 100 billion of them. The cerebellum at the lower rear has the majority, and the cortex has some 20 billion. But they are not the brain's only kinds of cells. Nerve cells are so delicate and specialized that they need help and support – which they get from glial cells. Glial (meaning 'glue') cells outnumber nerve cells by perhaps 20 to one, and they do a lot more than simply hold things together. Types of glial cell include astrocytes, oligodendrocytes and microglial cells.

ASTROCYTES

These cells support nerve cells both physically and with energy, nutrients and other needs; maintain and affect synapses; help the blood-brain barrier; and repair nerve and other glial cells.

OLIGODENDROCYTES

These cells make the fatty covering or myelin sheath of axons (see page 152); and support nerve cells physically and with nutrients.

250,000 million

MICROGLIAL CELLS

They are specialized 'resident defenders'. Like white blood cells, they scavenge for and remove invaders, damaged parts of brain cells and other unwanted material.

Speedy!

Microglia are the fastest-moving cells in the brain (apart from those borne along in a flow, as in blood). They race along at 0.1 mm per hour. At this rate they would take four days to cover 1 cm. Their extensions can lengthen or shorten twice as quickly.

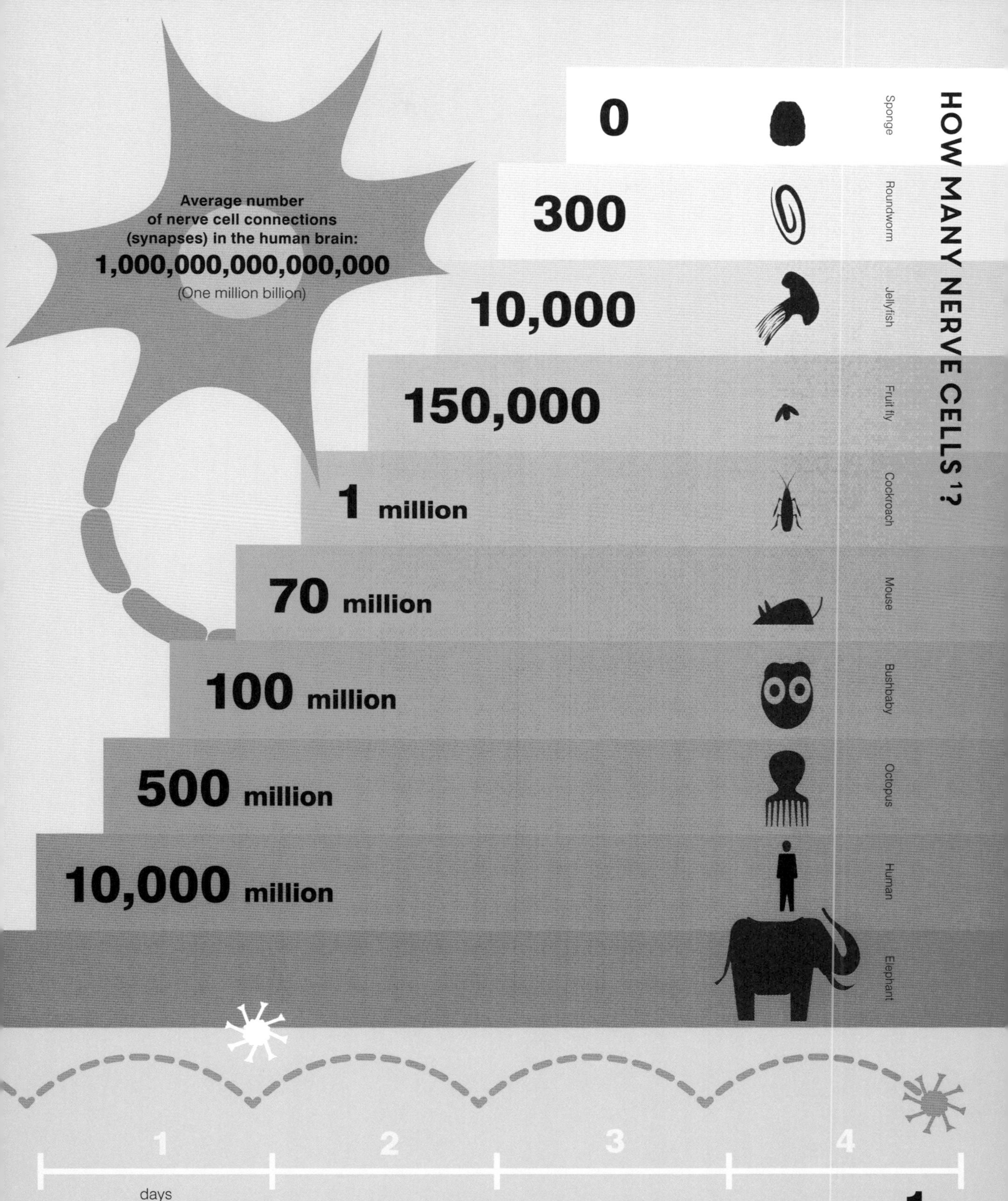

1 Neurons in the whole nervous system.

THE UNDERBRAIN

Beneath the great wrinkled dome of the right and left hemispheres, and in front of the 'mini-brain' cerebellum, are the mid brain, brain stem and other unfamiliar parts. They constantly work to keep the body's automatic systems running smoothly, to relay information between the consciously aware centres of the high brain and the rest of the body – as well as to carry out their own secretive, mysterious tasks.

Nucleus ruber (meaning 'Red body')
Some automatic aspects of movement, such as swinging arms when walking and running.

Substantia nigra (meaning 'Black stuff')
Part of mid brain. Planning and executing movements, head-eye movement coordination, pleasure and reward-seeking and addictive behaviour.

Tectum (meaning 'Roof')
This part of the mid brain deals with processing sight and sound information and eye movement.

Pons (meaning 'Bridge')
This link between lower and higher brain parts is involved in many and varied processes such as breathing, basic reflexes like swallowing and urination, sight and other main senses, facial movements, sleep and dreams.

Cerebellum (meaning 'Little brain')
Major hub for movement, balance and coordination.

Medulla (meaning 'Centre, core')
Also called medulla oblongata, this merges below into the spinal cord, and is involved with many automatic (autonomic or involuntary) processes, actions and reflexes including heart rate, breathing rate, blood pressure, digestive activity, sneezing, coughing, swallowing and vomiting.

BIG HEADED

BIGGEST BRAIN

Bigger creatures, in general, have bigger brains. But they are not always cleverer, at least by our measures of intelligence. A sperm whale cannot play chess or memorize the Sun's planets. (But then, a human cannot hunt giant squid a kilometre down in the ocean.) Measurments shows weight in grams.

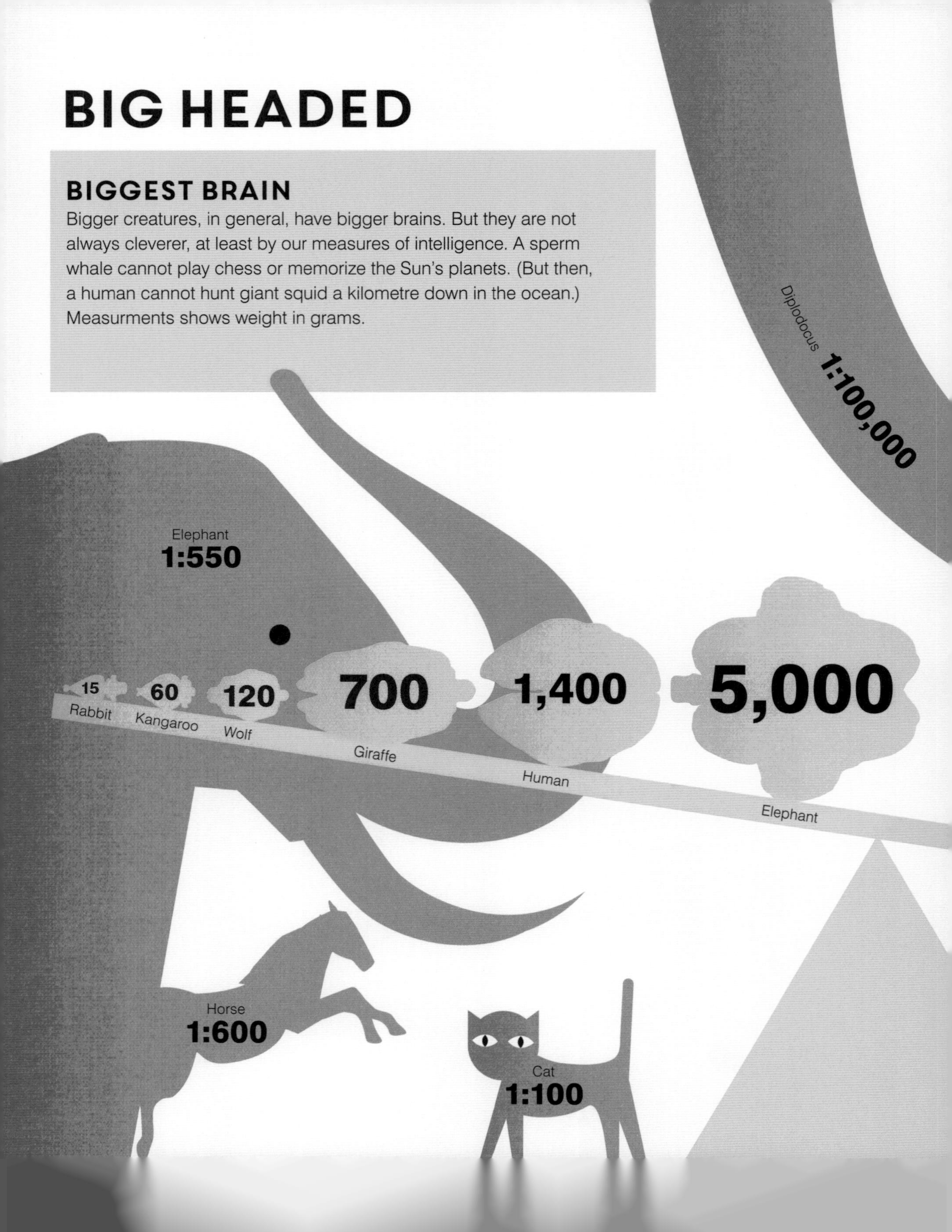

BIGGEST BRAINS FOR BODY SIZE

Comparing brain size to body size gives another measure that seems more linked to intelligence. Animals with larger ratios show features such as planning, problem-solving and adapting behaviour to novel situations. Ratio of brain mass to body mass.

Dolphin
1:100

Sparrow
1:15

Shark
1:2,500

Tree shrew
1:10

7,500

Sperm whale

Human
1:40

Ant
1:7

SENSES CROSS-OVER

The brain normally processes the main senses separately. But sometimes they get mixed or joined together. It can happen to anyone briefly, for example, a certain sound triggers a particular taste in the mouth, or a specific smell evokes mind-eye memories from long ago. In some people this sensory fusion is more common and known as synaesthesia. Words have colours (even when printed black), shapes provoke tastes, certain kinds of touch stimulate sounds.

% SYNAESTHETIC PEOPLE[1] COMBINATION EXPERIENCED

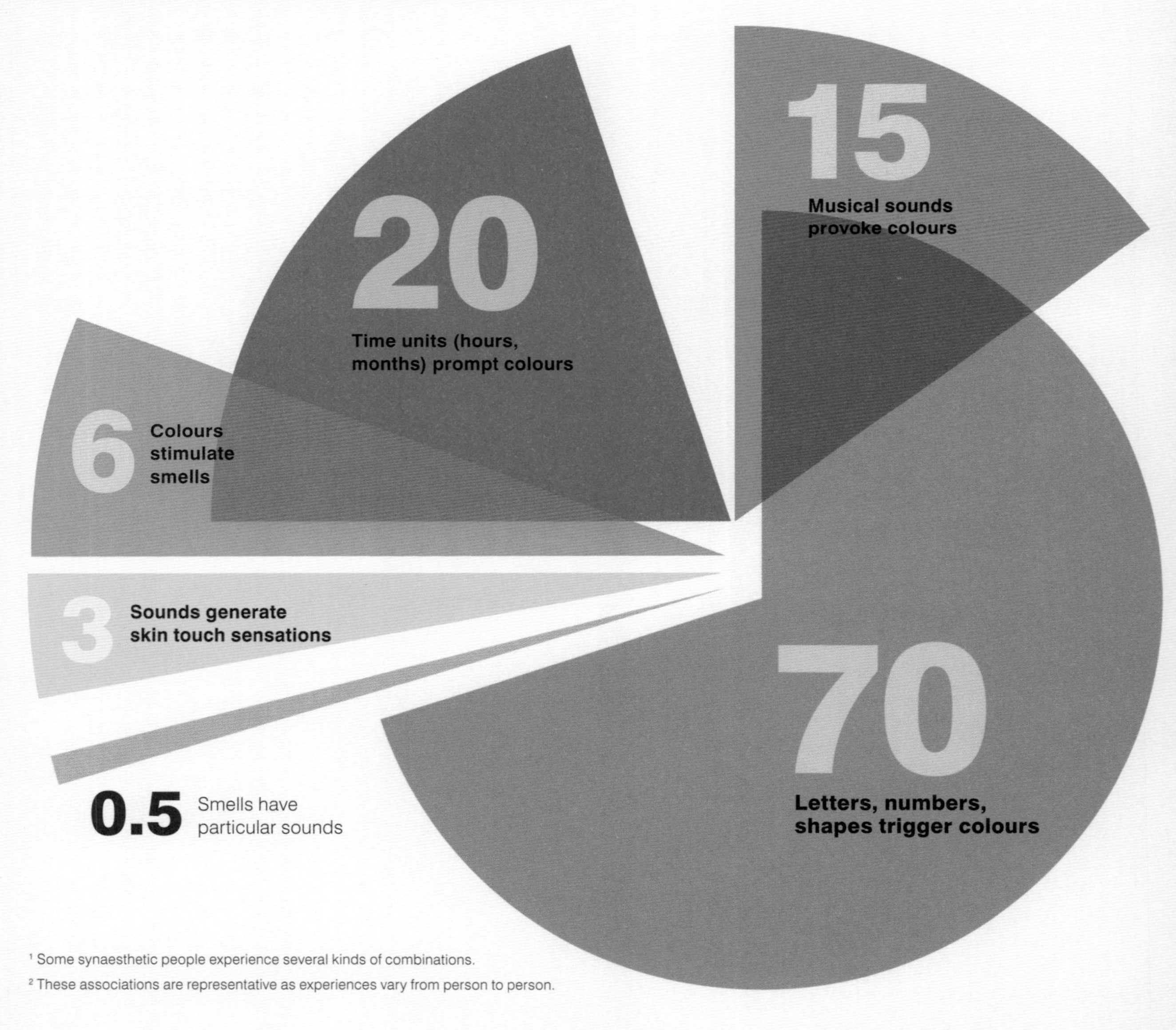

[1] Some synaesthetic people experience several kinds of combinations.

[2] These associations are representative as experiences vary from person to person.

TASTE-SOUND COMBINATIONS[2]

In some cases of synaesthesia a sound can trigger a particular taste in the mouth.

COLOURED MONTHS[2]

In other cases a month is associated with a colour.

MEMORY BY NUMBERS

Memory is immense. The brain holds not only facts and information, like a friend's phone number or who wrote *On the Origin of Species*[1]. It also remembers faces, scenes, sounds, smells, touches on the skin, skills and movement patterns like writing and cycling, plus experienced emotions and feelings. There are over-simple comparisons for the storage capacity of brains and computers. But vital, too, are the size of 'working memory' (in computers, RAM), and the speeds of storing and retrieving information.

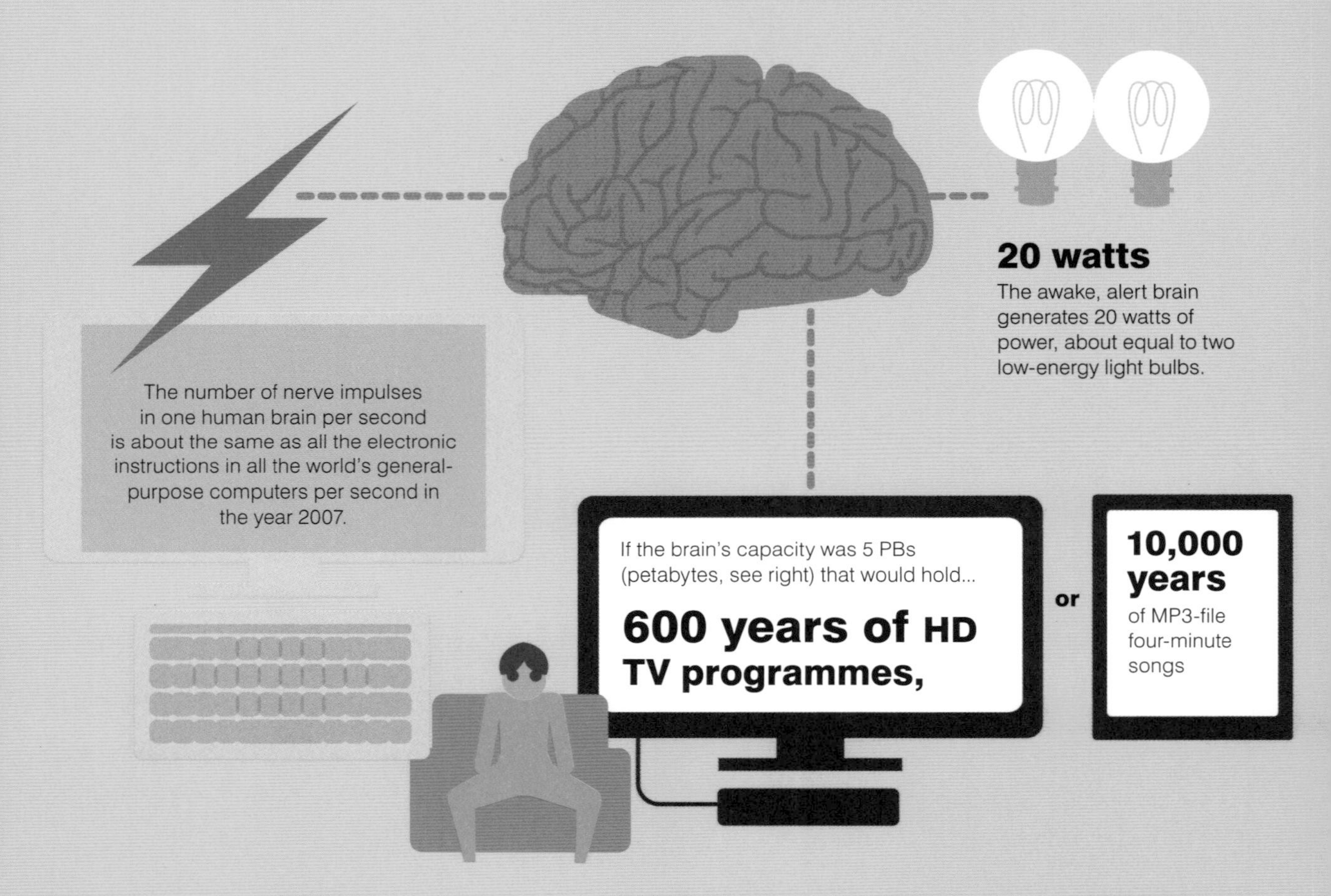

HOW FAST IS THE BRAIN?

One measure of computer processing speed or performance is FLOPS, floating point operations per second. A FLOP can be imagined as one calculation or computation. Assume:

- The brain has 100 million nerve cells.
- Each nerve cell connects to an average of 1,000 others.
- Each synapse, or connection between nerve cells, has about 20 different forms.
- Neurons fire up to 200 times per second.

Multiplied out, the brain's rate is 400 petaFLOPS (one million billion or one quadrillion FLOPS). This compares to supercomputer speeds of 10–50 PFLOPS.

1 Charles Darwin, book published 1859. Remember that.

HOW MUCH MEMORY?

Typical storage capacity of everyday working device

1
Home computer hard drive

150
1 A4 page word document

100–200
TV HD hard disc recorder

8–64
Memory stick

16–64
Tablet or smartphone

1 brain synapse
0.0047

10–100
Supercomputer

10–100
Human brain lower estimates

1–10
Human brain upper estimates

B: Byte	Usually 8 bits, 1 working unit of memory	
KB: Kilobyte	1,000 bytes	
MB: Megabyte	1,000 KB	1 million bytes
GB: Gigabyte	1,000 MB	1 billion bytes
TB: Terabyte	1,000 GB	1 trillion bytes
PB: Petabyte	1,000 TB	1 quadrillion bytes

THE MEMORY GAME

Perhaps inconveniently, the brain has no single 'memory centre'. Indeed, there is no single type of memory, but several kinds. Numerous brain parts handle different aspects of their learning, storage and recall. These parts are also wired into other brain regions, including emotional areas. So moods and emotional states, along with fatigue, hunger, distractions and many other factors, greatly affect memory. At the level of cells, a memory is a new pattern of connections and pathways between the brain's billions of neurons.

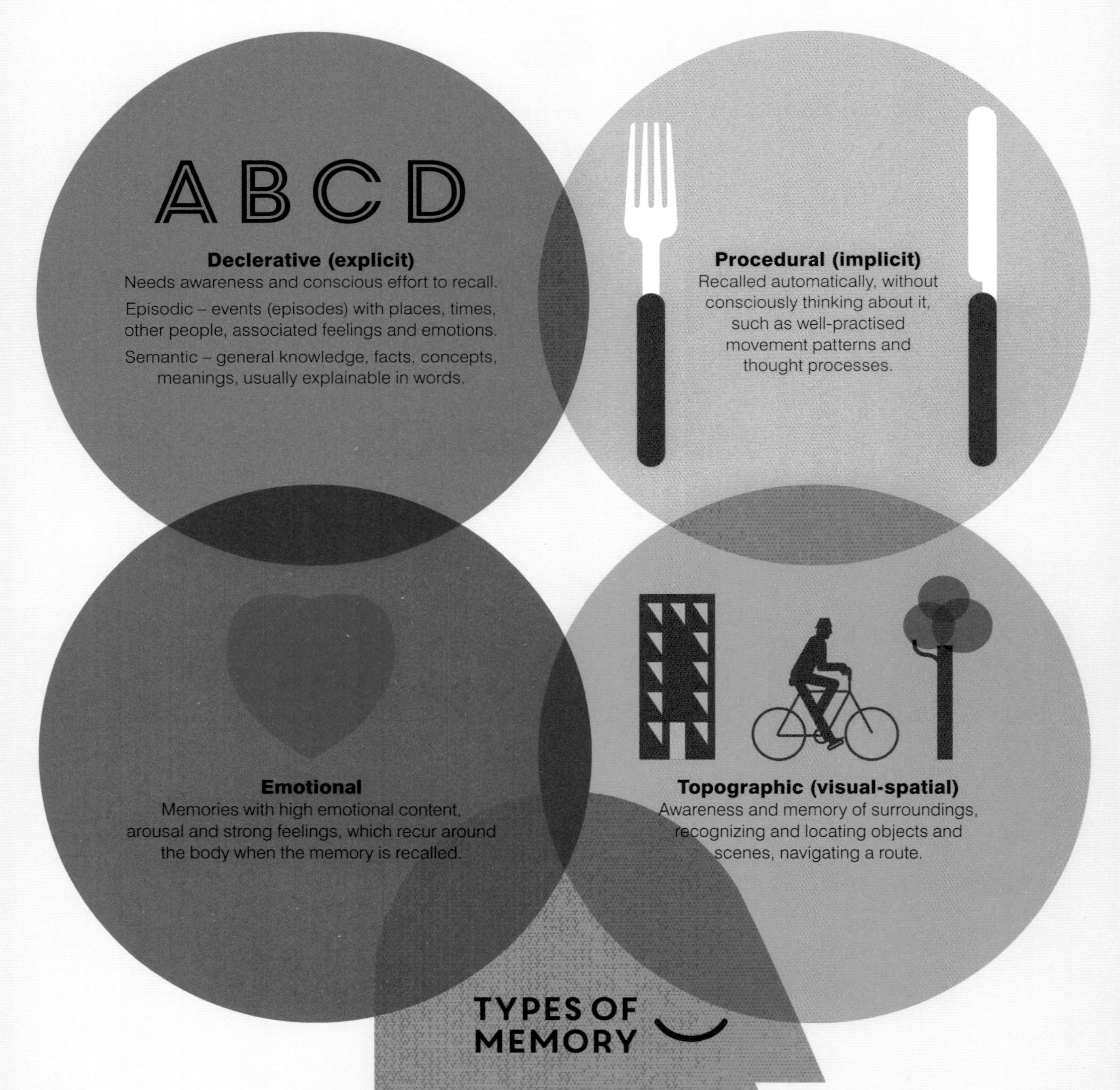

Movement (motor) cortex
Holds movement memories (procedural).

Touch cortex
(Somatosensory)
Stores touch memories.

Frontal lobes
Main site of short-term 'working memory' such as topographic awareness. Hold much associative information linking to other regions for various elements of a memory.

Hearing cortex
Holds sound memories.

Taste cortex
Holds flavour memories.

Smell cortex
Stores odour memories.

Amygdala
Major roles in forming memories with high content of emotions and feelings (emotional memory). Important roles in memory consolidation, converting short- to long-term memory (with hippocampus).

Hippocampus
Important roles in memory consolidation, converting short- to long-term memory (with amygdala). Involved in spatial memory of objects in surroundings and navigation (topographic memory).

Sight cortex
Stores visual memories.

Cerebellum
Stores movement memories (procedural).

SHARING MEMORIES

Many brain parts hold different aspects or components of memory. For instance, the vision centre or cortex holds the image-based information that allows an object to be recognized, named and incorporated into a larger memory experience. Much combining of memory elements into conscious awareness happens in the frontal lobes.

EMOTIONAL BRAIN

'Is that true? Oh no, how awful. Tragedy!' The body responds with faintness, trembling, unsteadiness, perhaps sobbing. The mood is distraught, the mind cannot think straight or make sensible decisions. 'No, wait – it's not true. Fantastic!' Spirits rise, the body jumps for joy. Cries of anguish turn to delight, and tears of pleasure replace those of pain. Where in the brain do such powerful emotions come from?

THE LIMBIC SYSTEM

This system is defined by function, that is, parts that contribute to feelings, moods and emotions. (The parts have various other tasks too.)

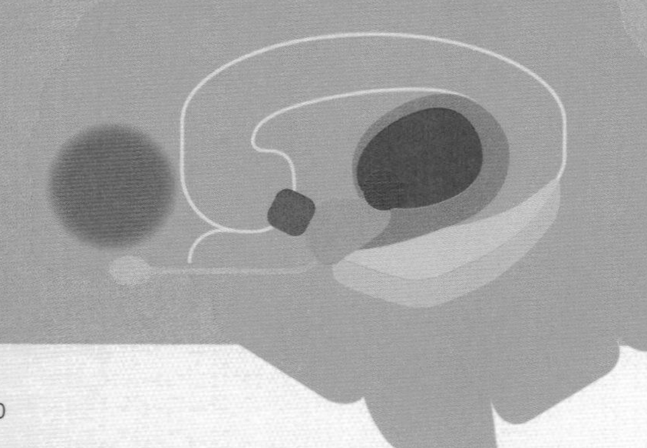

Hippocampus
Forms and exports long-term memories (but does not store them itself). With the amygdala, helps in emotional components of memory and their recall.

Amygdala
Highly active (with hippocampus) in processing memories and their recall. Especially involved in emotions, both minute by minute, recalled, and even imagined.

Olfactory bulb
Sends smell messages direct to amygdala, hippocampus and other limbic parts. This is why odours and perfumes evoke such strong, immediate emotions and forceful memories.

WHERE ARE EMOTIONS FELT?

Each person has subjective feelings for which parts of the body are affected by strong emotional states. The feelings can be mapped onto a body outline.

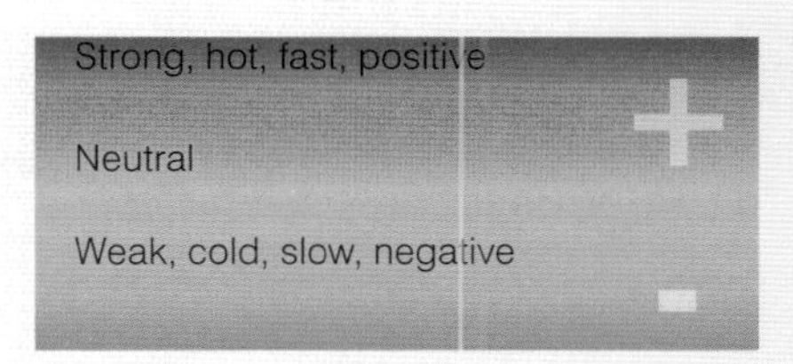

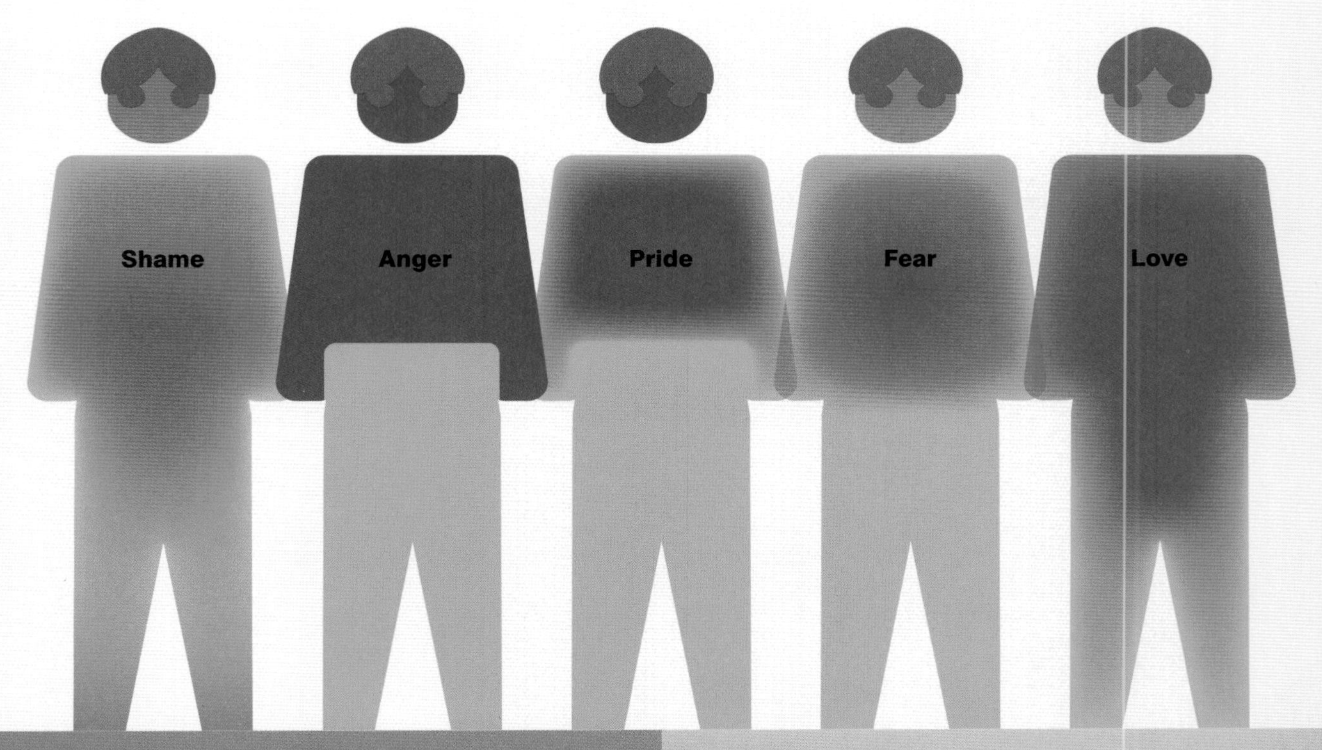

Fornix
Intermediary between hippocampus, thalamus and mammillary bodies. Contributes to emotional aspects of memory.

Mamillary bodies
Involved in episodic memories, those dealing with events (episodes): places, times, people, feelings.

Thalamus
Relay station and distribution centre for other limbic parts.

Parahippocampal gyrus
Memory and recognition of whole scenes (rather than people, objects in them) and emotional responses to these.

Hypothalamus
Tends to be involved in bodily expression of emotions rather than generating or originating them. Linked with emotional states such as disgust, displeasure, and uncontrollable laughter and tears.

Limbic area of frontal lobe
Front, lower, inward-facing region of the brain surface. Major hub and association area for many kinds of memory, including spatial awareness and navigation. Transfer area between hippocampus and its linked areas, and the rest of the cortex.

BRAIN TIME

The body has its own built-in biological clock, the SCN, suprachiasmatic nucleus. Its nerve cells have a 24-hour or circadian ('about one day') activity cycle of their own – well, approximately. This activity is synchronized or entrained to the outside world by the natural daylight-darkness sequence detected by the eyes, which 'sets' the clock. The SCN controls and coordinates biorhythms across the board, from body temperature and hormone levels to appetite, digestion, waste removal and the wake-sleep cycle.

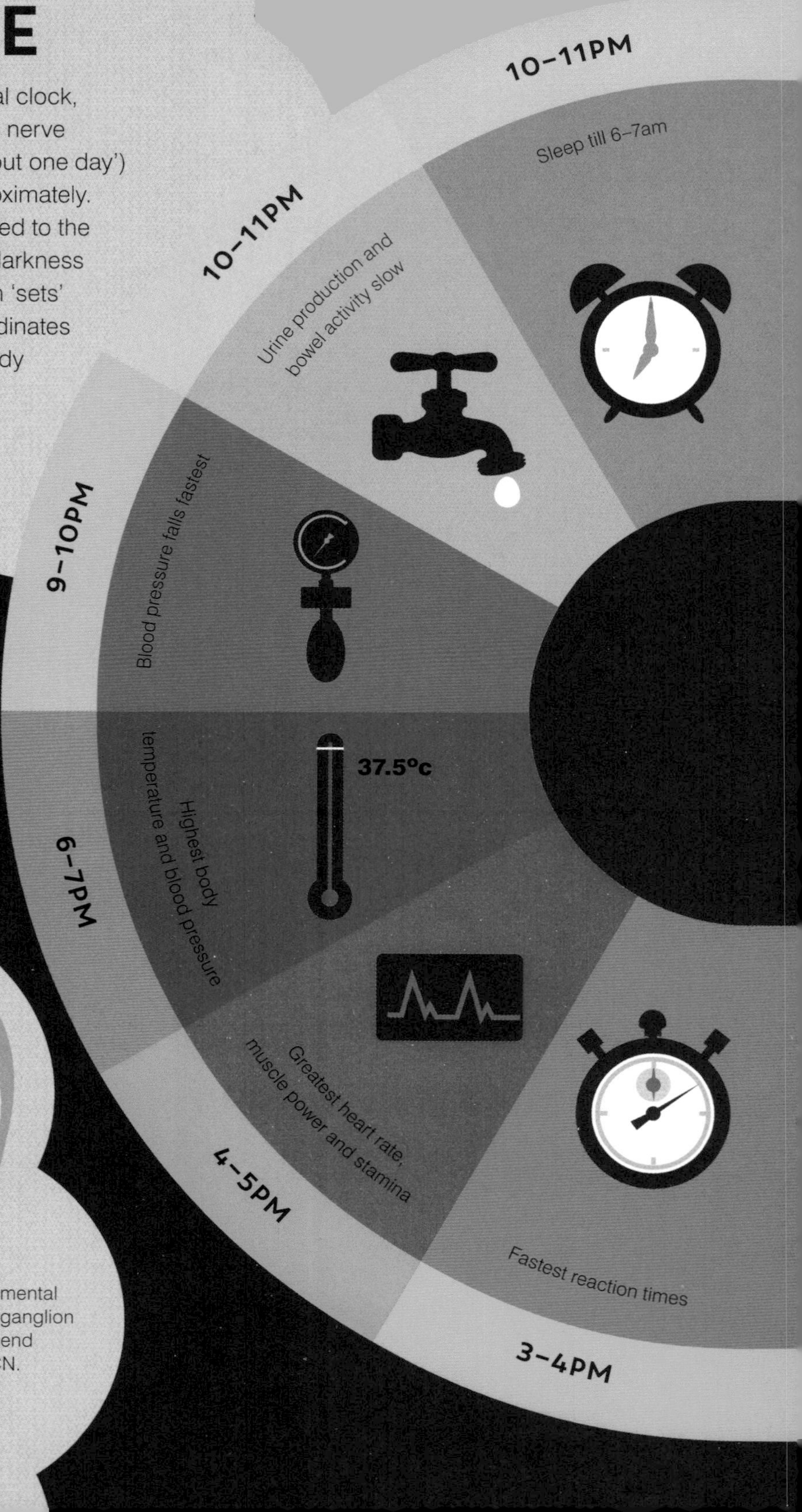

10–11PM
Sleep till 6–7am

10–11PM
Urine production and bowel activity slow

9–10PM
Blood pressure falls fastest

6–7PM
Highest body temperature and blood pressure
37.5°c

4–5PM
Greatest heart rate, muscle power and stamina

3–4PM
Fastest reaction times

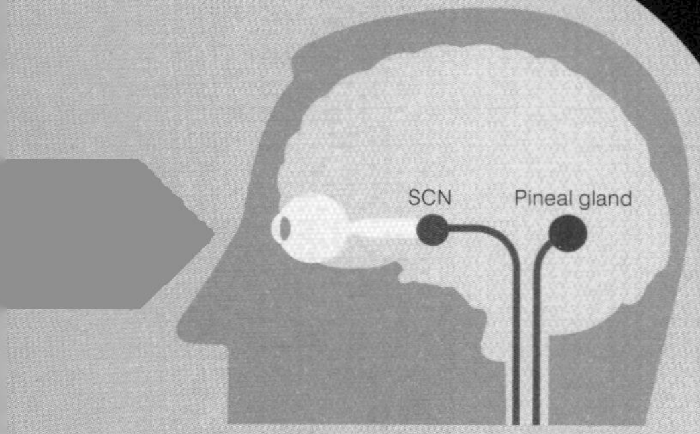

SCN
Pineal gland

Clock-setting: 1
Daylight time is the primary environmental signal. Light levels are detected by ganglion cells in the retina of the eye which send messages almost directly to the SCN. Other cues are shown opposite.

4–5AM

Lowest body temperature

36°c

7AM

Wake up.
Blood pressure rises fastest

7–8AM

Bowel movements
and urination likely

X+Y

10–11AM

High levels of mental alertness

12–1PM

Keenest appetite

2–3PM

High levels of physical
coordination, high pain threshold

Clock-setting: 2

Ambient temperature is another external change, and is detected by the skin. Incoming food and mealtimes are monitored by various brain parts such as PBN, parabrachial nucleus. Stress raises the level of the stress hormone cortisol, and exercise raises body temperature, also heart and breathing rates.

The body's daily rhythms follow a circadian clock and involve many parts of the hormonal or endocrine system, especially the pineal gland.

OFF TO SLEEP

Consolidating important and well-used memories and discarding insignificant and little-used memories.

Energy use and general metabolism.

Wound healing

Intestines and digestive activity.

Heart rate and blood pressure.

'Rewiring' nerve cell connections throughout brain to reinforce learning

Cortex activity in brain.

Breathing rate.

Immune system activity.

Kid
urine

Speeds up

Slows down

HO

C
Me

6am
6pm
6am

We spend one-third of our lives asleep, mainly due to melatonin from the brain's pineal gland. Sleep happens in stages from light or shallow to deep, plus a renegade stage called REM when dreams happen. EEGs, electroencephalograms, are recordings of the brain's electrical activity, tracking the numbers, locations and patterns of nerve signals. Each sleep stage and major mental process has a characteristic EEG trace. Through all this, the brain certainly does not rest; it's especially busy processing memories. Subsistence organs like the heart, lungs, guts and kidneys take it easy. The immune and tissue-maintenance systems up their game and rush ahead with their tasks.

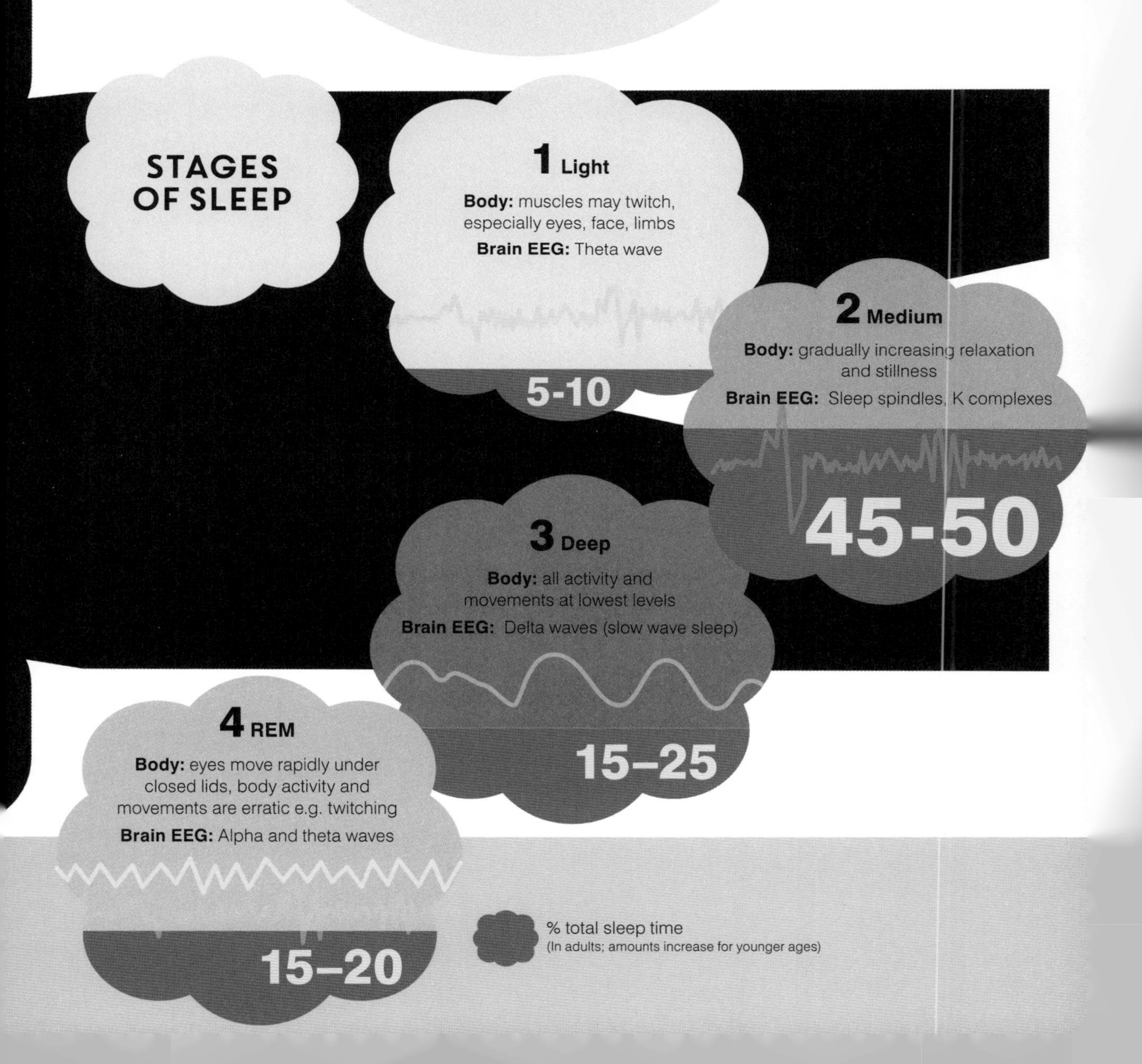

NEED FOR REM SLEEP

Sleep needs vary hugely between individuals.
Sufficient REM sleep is especially important for good health.

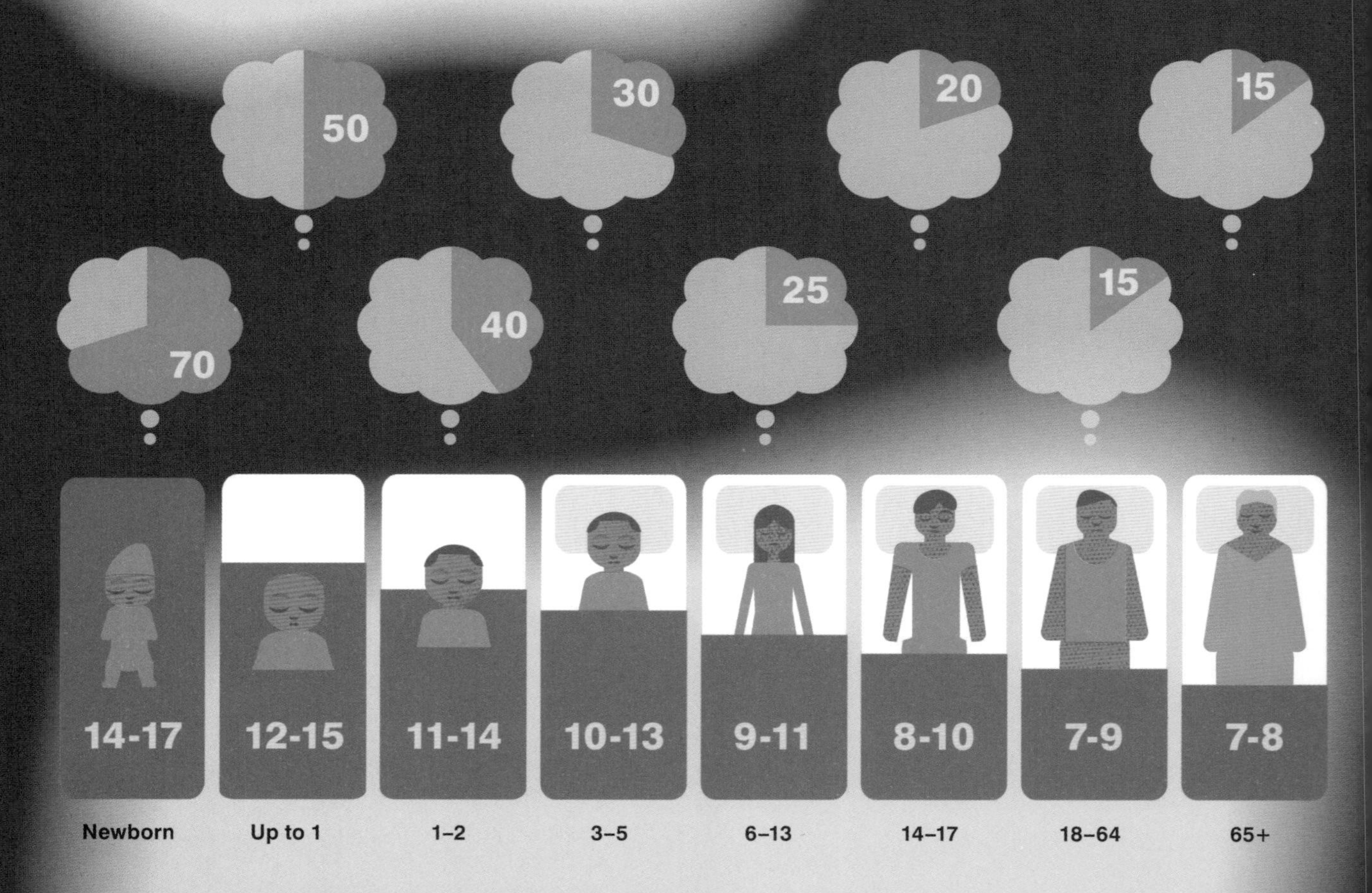

DREAMTIME

When people are wired up for EEGs and other body functions during sleep testing, and then woken up during REM – they usually report that they have been dreaming. Dreams can be reassuring, peculiar, unsettling or truly nightmarish. EEGs and scans reveal which parts of the brain are engaged. However, the serious science of interpreting dreams has a long way to go.

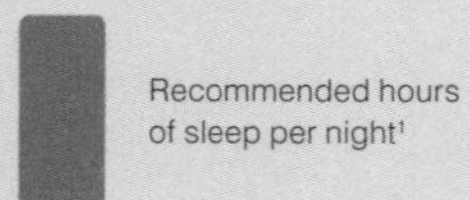

1 Guidelines from US National Sleep Foundation

WHERE'S BUSY DURING SLEEP?

Sleepy teens
It's official: adolescents really do find it difficult to get up in the morning. Research shows the body clock and daily biorhythms tend to run one or two hours later during the teenage years.

QUIET

- 1 **Motor (movement) centre**
- 2 **Touch centre**
- 3 **Primary visual (sight) centre**
- 4 **Hearing centre**
- 5 **Frontal lobe:** dampens conscious inputs

ACTIVE

- 6 **Smell centre:** strong odours may wake dreamer
- 7 **Associated visual areas:** dream imagery
- 8 **Thalamus:** filters many sensory inputs to cortex
- 9 **Amygdala:** memory links to emotions
- 10 **Hippocampus:** short-term in-dream memory loss
- 11 **Medulla:** basic life maintenance

GROWING BODY

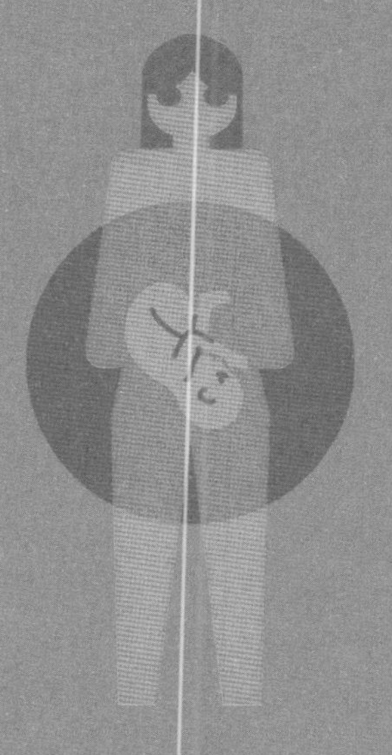

A rule of life is 'All cells come from other cells' by division, or mitosis. A new life is the same, but with complications. Each cell in the body has a double-set of genetic material. Babies are made from an egg cell and a sperm cell. If these two had a double set each, the result would be a quadruple set. So the double set must be halved to single, then sperm and egg fuse to make a twin set to start the new baby. This is where a special kind of cell division comes in – meiosis, to make eggs and sperm.

GAMETE FORMATION IN THE MALE

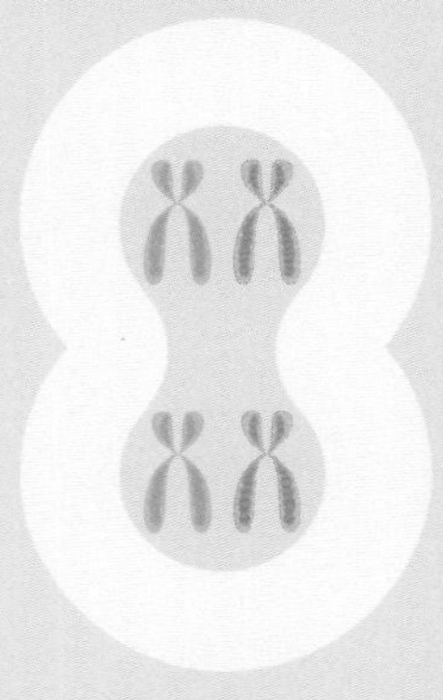

Interphase

DNA forming pairs of chromosomes replicates. Result is two sets of 23 pairs of chromosomes.

Prophase/Metaphase 1

Chromosomes become visible. Some chromosomes may swap sections with their partners (crossing over) to introduce genetic variation. Nuclear membrane disintegrates. Chromosomes align at centre or equator of cell.

Anaphase/Telophase 1

Pairs of chromosomes separate, one pair into each new cell. Nuclear membrane reforms in each sibling cell. Original cell has divided into two, each with a pair of each chromosome

GAMETE FORMATION IN THE FEMALE

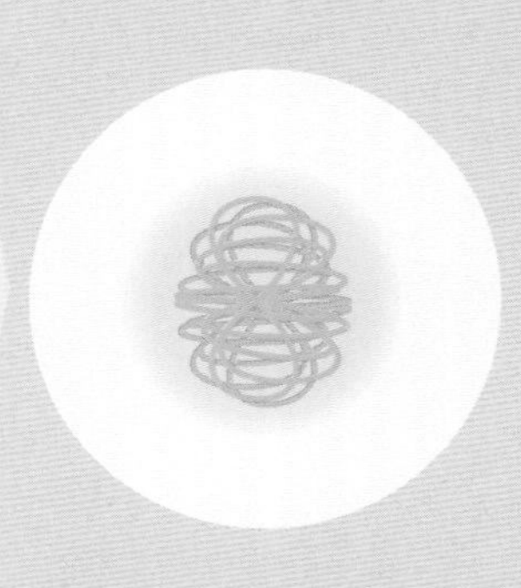

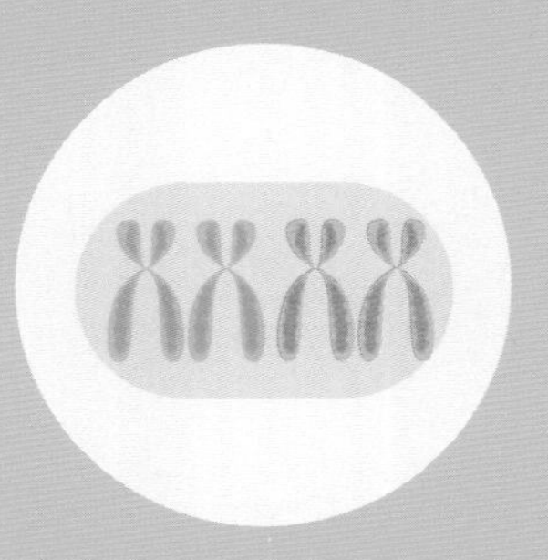

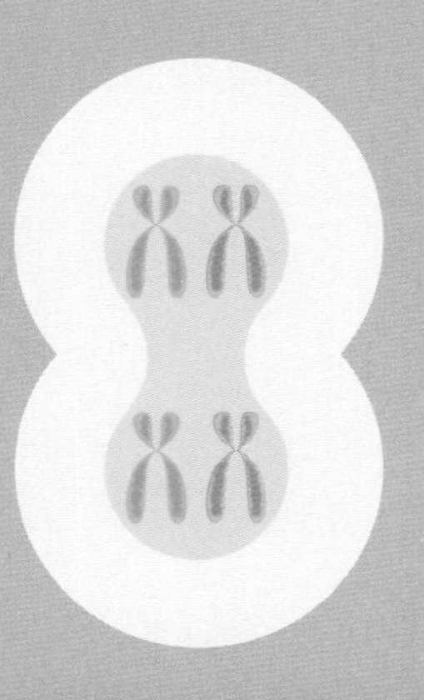

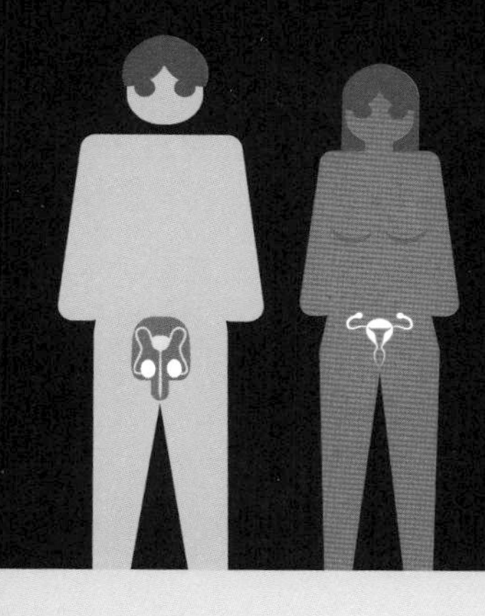

The male gamete is called a sperm and contains 23 chromosomes, half the number needed for the formation of a zygote, the first cell of a new individual.

The female gamete is called an ovum (or egg) and contains 23 chromosomes, half the number needed for the formation of a zygote, the first cell of a new individual.

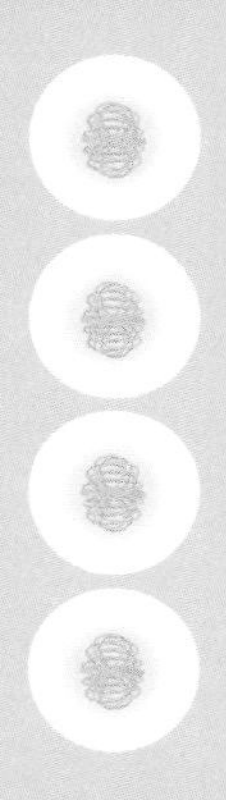

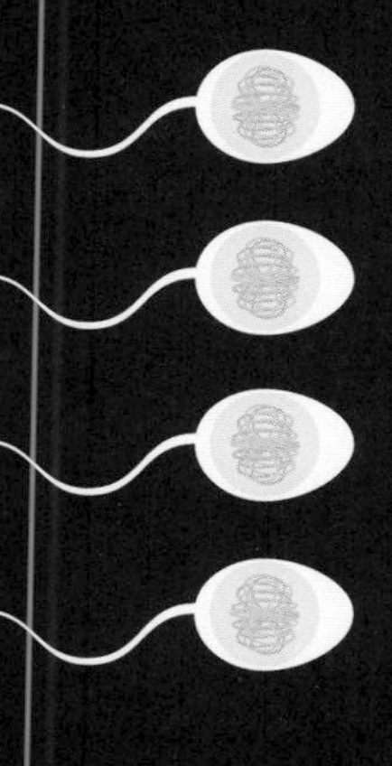

Prophase/Metaphase 2

Nuclear membrane disintegrates. Chromosomes align randomly at centre or equator of cell.

Anaphase/Telophase 2

Chromosome pair separates, one into each new cell. Nuclear membrane reforms in each sibling cell.

Original cell has divided into four, each with only one of each chromosome

One original male cell produces four sperm cells. One original female cell produces one egg cell plus three polar bodies (containing 'spare' chromosomes).

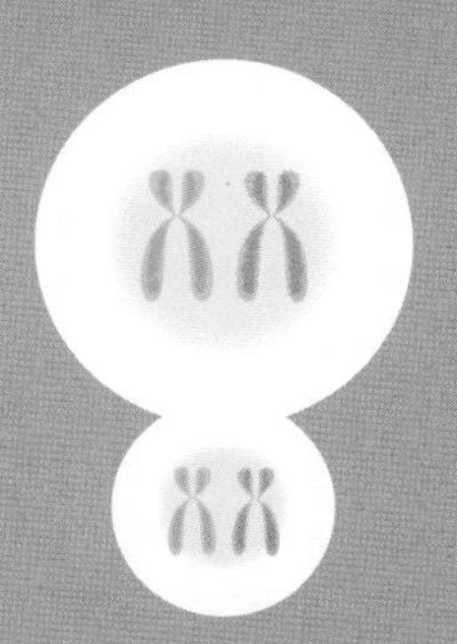

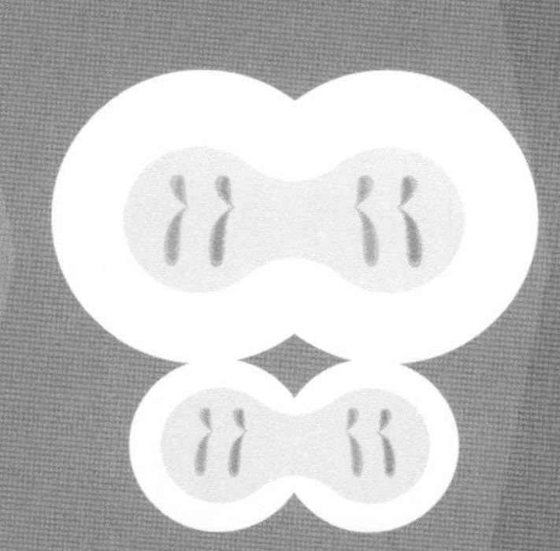

MAKING EGGS

When the two sex cells – egg and sperm – join to start a new baby, they contribute an equal share of genes. Each has 23 chromosomes, each chromosome made of one length of DNA. But production of mature sex cells is very unequal. In the female it begins around puberty, occurs about once every 28 days during the monthly cycle, and halts at the menopause. Sperm production, in great contrast, is a 24/7 process and fades gradually with age.

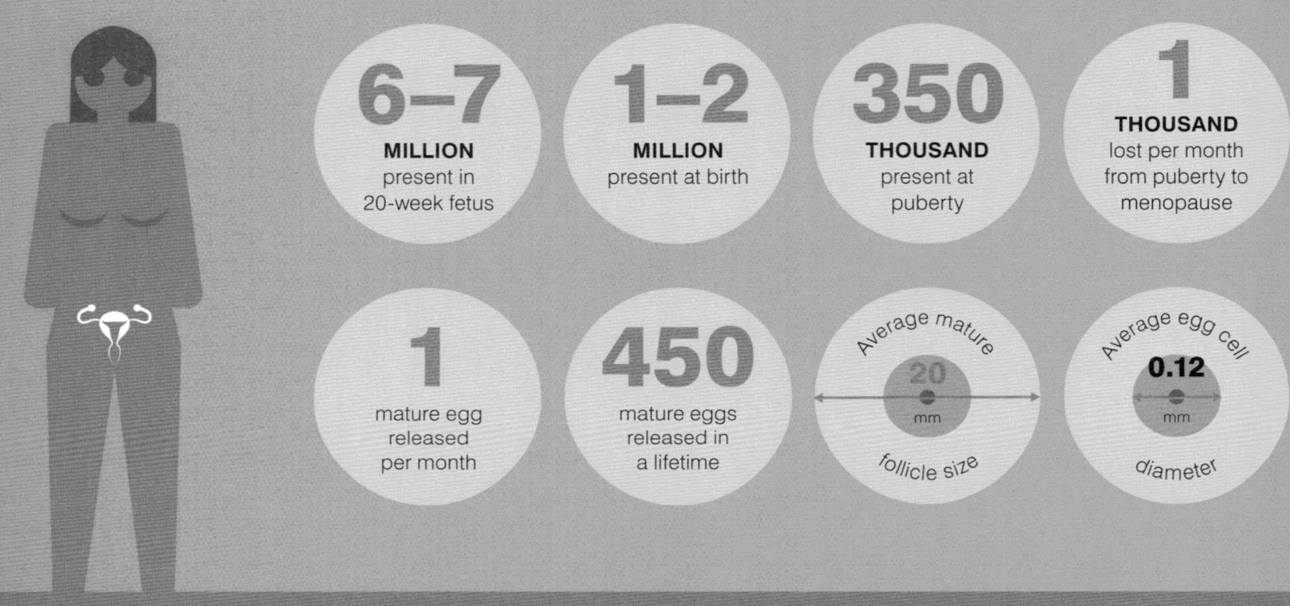

REPRODUCTIVE CYCLE

The female reproductive cycle is coordinated by several hormones including FSH, LH, oestrogen and progesterone

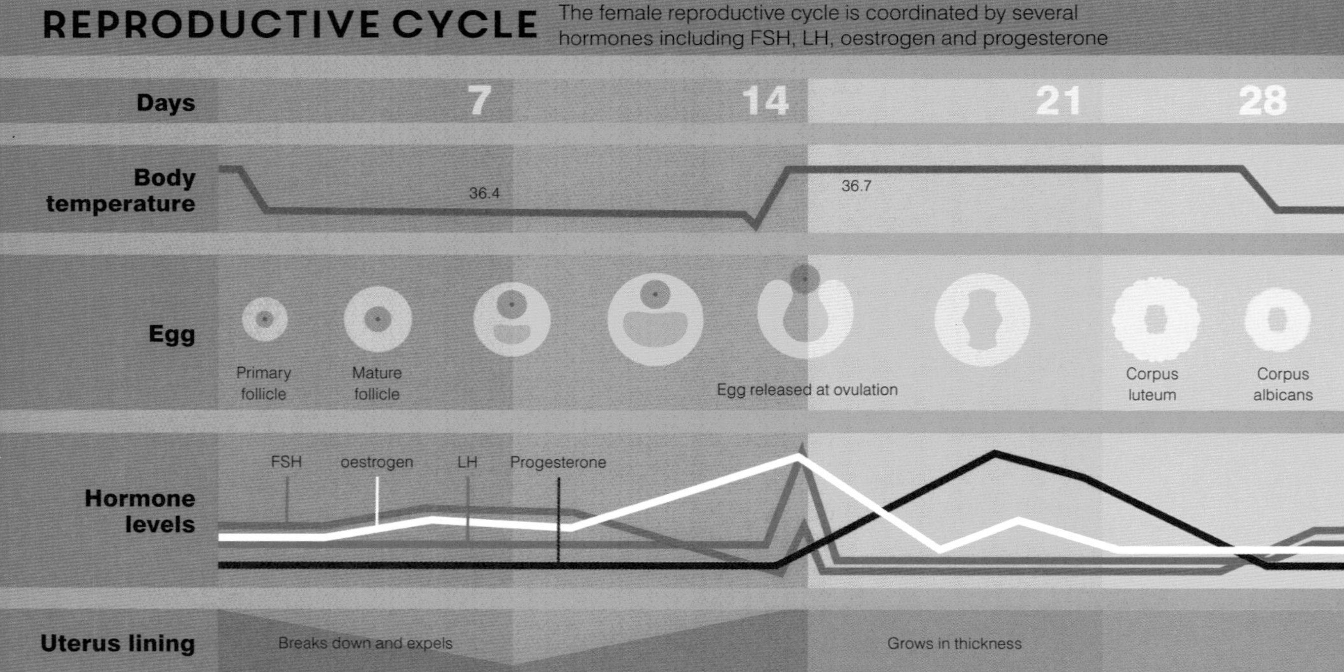

MAKING SPERM

Production of male sex cells, sperm, is a continuing process and happens in vast numbers, with millions daily growing and maturing within the testes. The production line begins around puberty, then continues every minute of every day, until it gradually tapers off with old age. However, men into their 70s and 80s have still fathered children in the natural way.

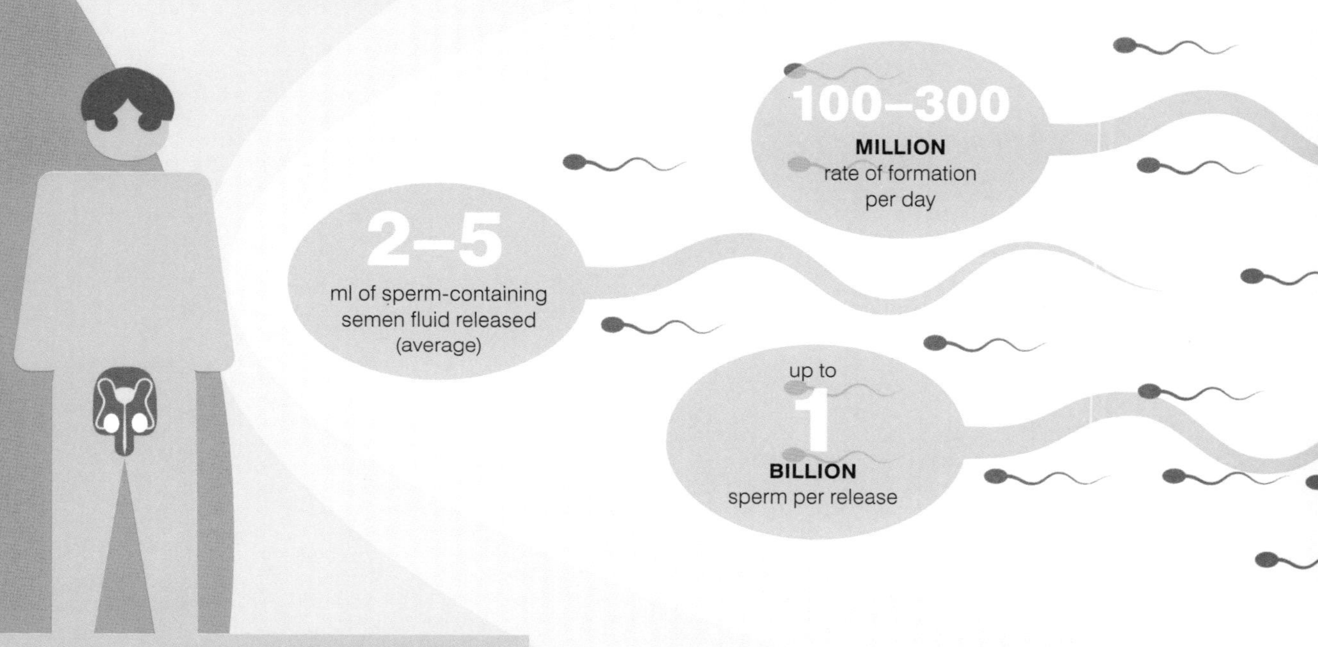

5
MILLION MILLION
mature sperm produced in a lifetime

SPERM PRODUCTION

Days spent in each stage of development

Stage	Days
Spermatogonia stem cells	8–10
Primary spermatocyte	12–15
Secondary spermatocyte	15–17
Round spermatid	18–22
Elongated spermatid	25–30
Spermatozoon	30–35
Total time taken to develop and mature	3+ months 3

NEW BODY BEGINS

The joining of egg and sperm cells to start a new baby is known as fertilization – and also conception, even syngamy. It usually happens in the oviduct (fallopian tube) leading from the ovary, where the egg came from, to the uterus (womb) where the baby will develop. The successful sperm is not just one in a million, but perhaps one in a billion. Nearly all of its fellows fail to reach the egg, and once it does make egg contact, the egg stops any more sperm joining. The joining of sperm and egg at conception sets off an amazing process of growth and development that, nine months later, results in a wrinkly, wailing, small human.

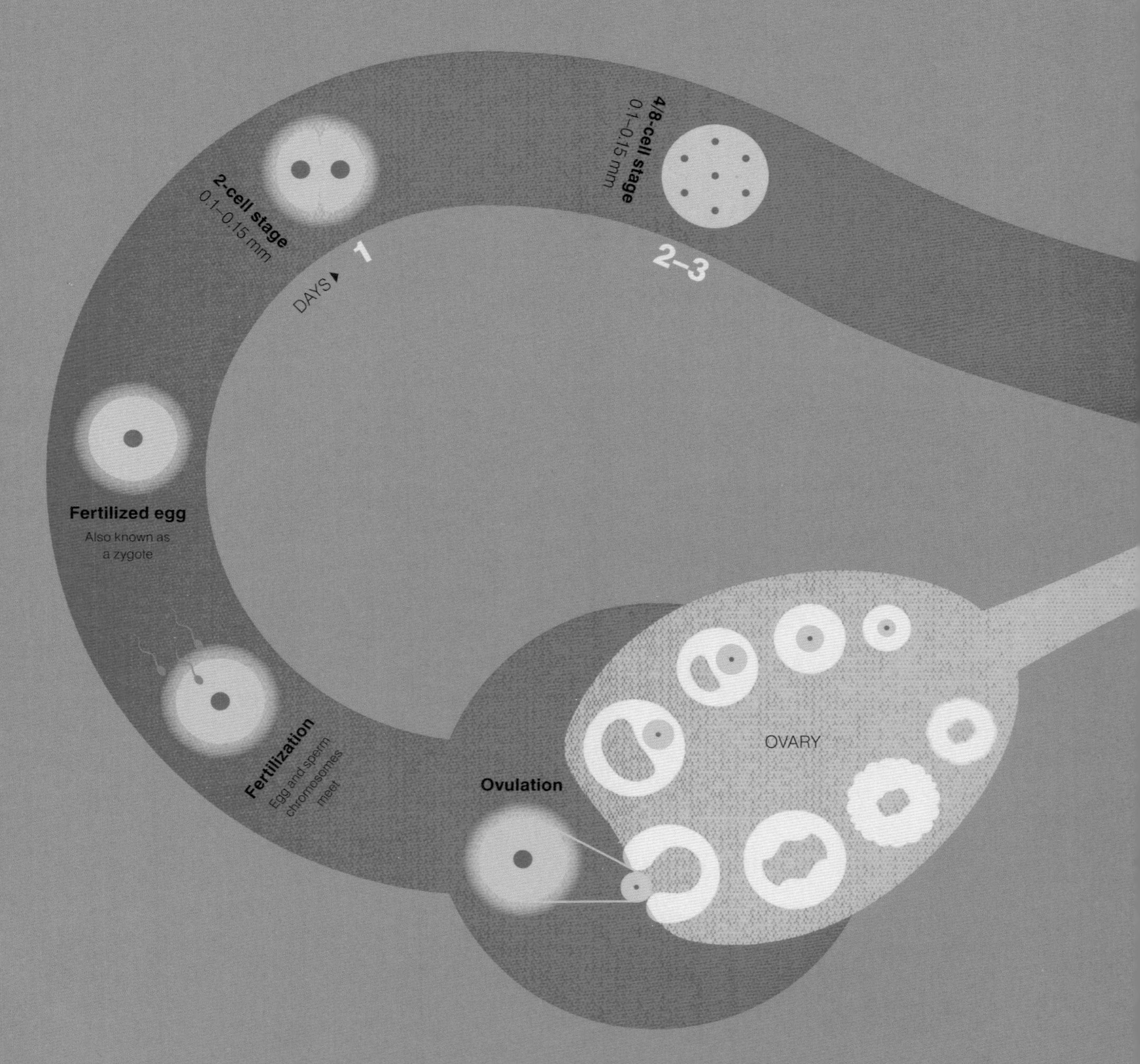

STAGES OF FERTILIZATION

1. Only a few hundred sperm reach the egg
2. Many sperm try to make contact
3. Cap (acrosome) releases enzymes to dissolve zona and egg outer membrane
4. One sperm head fuses with egg outer membrane
5. Sperm chromosomes in nucleus pass into egg
6. Zona and outer membrane toughen to prevent more sperm fusing
7. Sperm and egg chromosomes gather together and fertilized egg prepares for first division

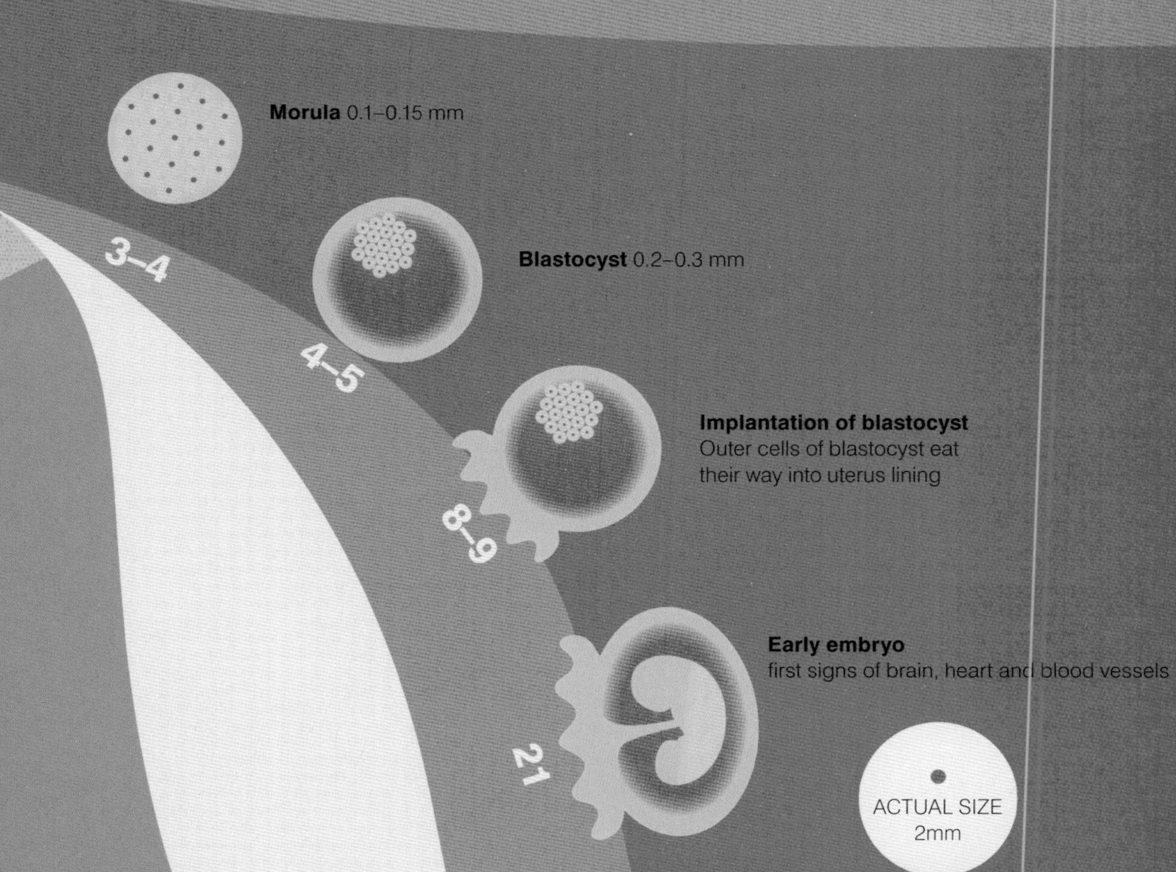

PREGNANCY TIMELINE

The baby grows in its very special place, the uterus. But it's not all calm, peace and serenity. The mother's heartbeats thud above and her blood whooshes through nearby arteries. Bright lights filter through the skin and uterus wall. So do sudden loud noises that can startle the baby so it punches or kicks. Increasing size means the baby becomes more cramped, and it is squeezed and squashed as the mother moves around.

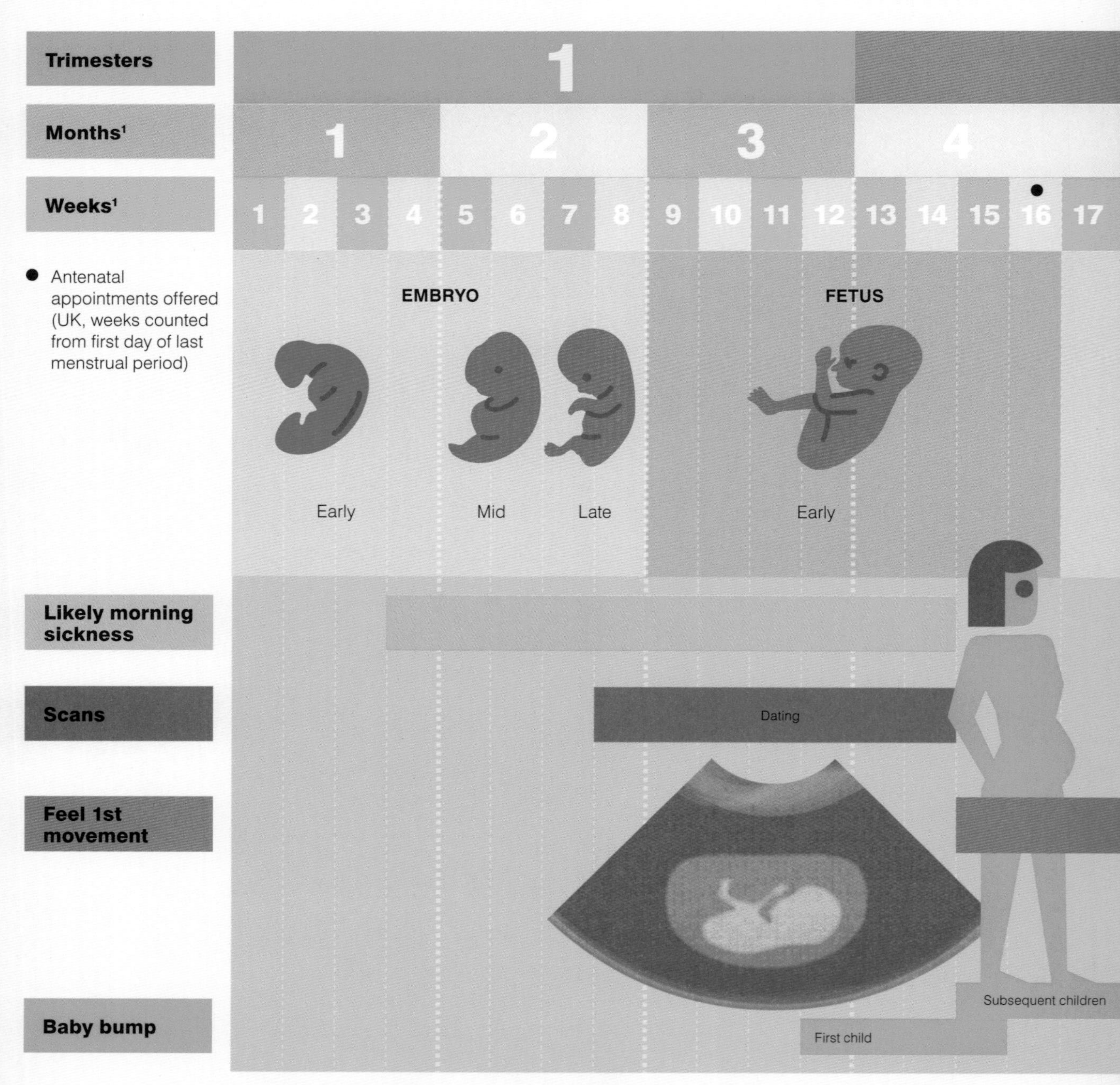

1 Times from fertilization of egg by sperm, or conception. Some timelines are taken from the date of the mother's last menstrual period two weeks earlier, giving a total of 40 weeks.

PREGNANCY TEST ACCURACY

Tests detect the hormone hCG in the mother's urine, produced about 6 days after conception.

Accuracy, %	60	90	97
Days after conception	10	14	18

2

3

5

7

9

18 19 20 21 22 23 24 25 26 27 28 29 30 31 32 33 34 35 36 37 38

FETUS

Mid

Viable[2]

Anomaly

2 Viability and perinatal period are defined in different ways by different authorities depending on, for example, improvements in newborn care and the proportion of babies likely to survive at a certain stage of development.

THE UNBORN BABY

Multiply, move, specialize: it happens every minute of the nine months before birth. The hundreds of cells in the embryo multiply to become thousands, then millions. They also physically move or migrate, forming folds, lumps and sheets that gradually shape the organs. And they differentiate, that is, they change from the general, all-purpose stem cells of the early stages, into particular different kinds such as bone, muscle, nerve and blood cells.

4

• Heart beats 120–140 times per minute •
• Eye spots on head •
• Muscles forming, some movements occur •
• Arm buds appear •
• Tail present •

4 mm

25 cm²

24

• Heart beats 150 times per minute •
• Head is one-quarter total length •
• Eyes can open •
• Thumb-sucking may occur •
• Early memories may form •

WEEKS[1]

8

• Facial features recognizable •
• Head as big as body •
• Fingers and toes forming •
• Tail shrinks •
• Name changes from embryo to fetus •

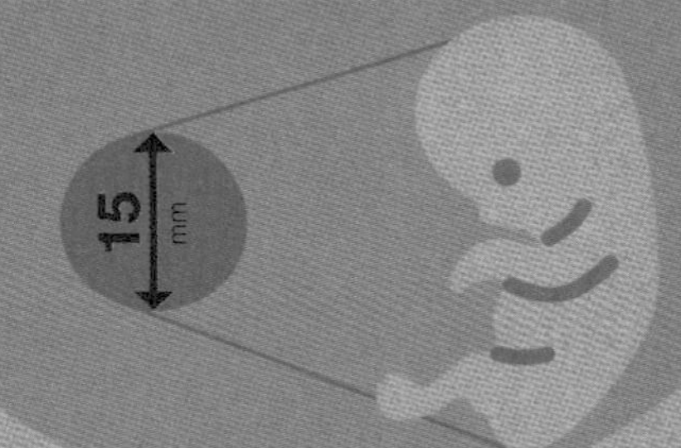

16

• Face recognizably human •
• All organs formed •
• Milk teeth buds present in jaws •
• All bones shapes present, although mostly as cartilage •
• Fat starts to accumulate under skin •

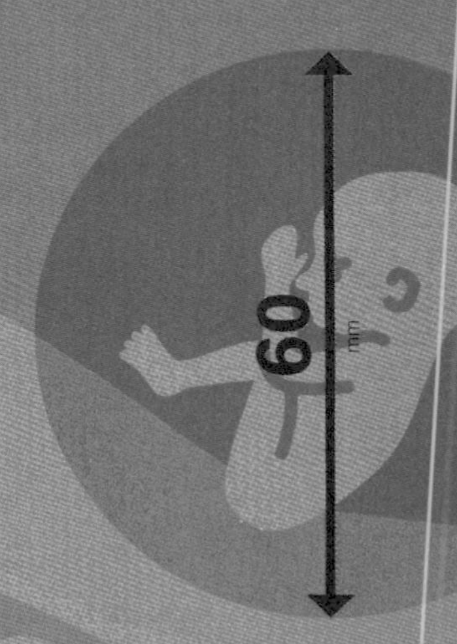

45–48 cm

36

• Lanugo (downy first hair) is shed •
• Nails may extend beyond fingers and toes •
• Coughs and hiccups common •
• Baby is ready for birth •
• Weight 3+ kg •

1 Times from fertilization of egg by sperm, or conception. Some timelines are taken from the date of the mother's last menstrual period two weeks earlier,giving a total of 40 weeks.

2 Because it is usually curled in the 'fetal' position, embryo/fetus length is usually crown-rump, from top of head to base of bottom.

BIRTH DAY

Time taken to give birth famously varies from less than an hour to 24-plus hours; it usually reduces by 30–40% for second babies, and perhaps another 10–20% thereafter. Birth statistics in developed nations are being impacted by more assisted, intervened and managed deliveries, especially induced labour and C-section (caesarean delivery). This means fewer babies are now born on Sunday that any other weekday, and the date with least births is often 25 December.

ORGAN GROWTH

Brain and eyes are huge in proportion in a baby compared to an adult. But even they are outdone by the thymus gland in the chest – already more than half its adult weight.
Organ as % of adult weight

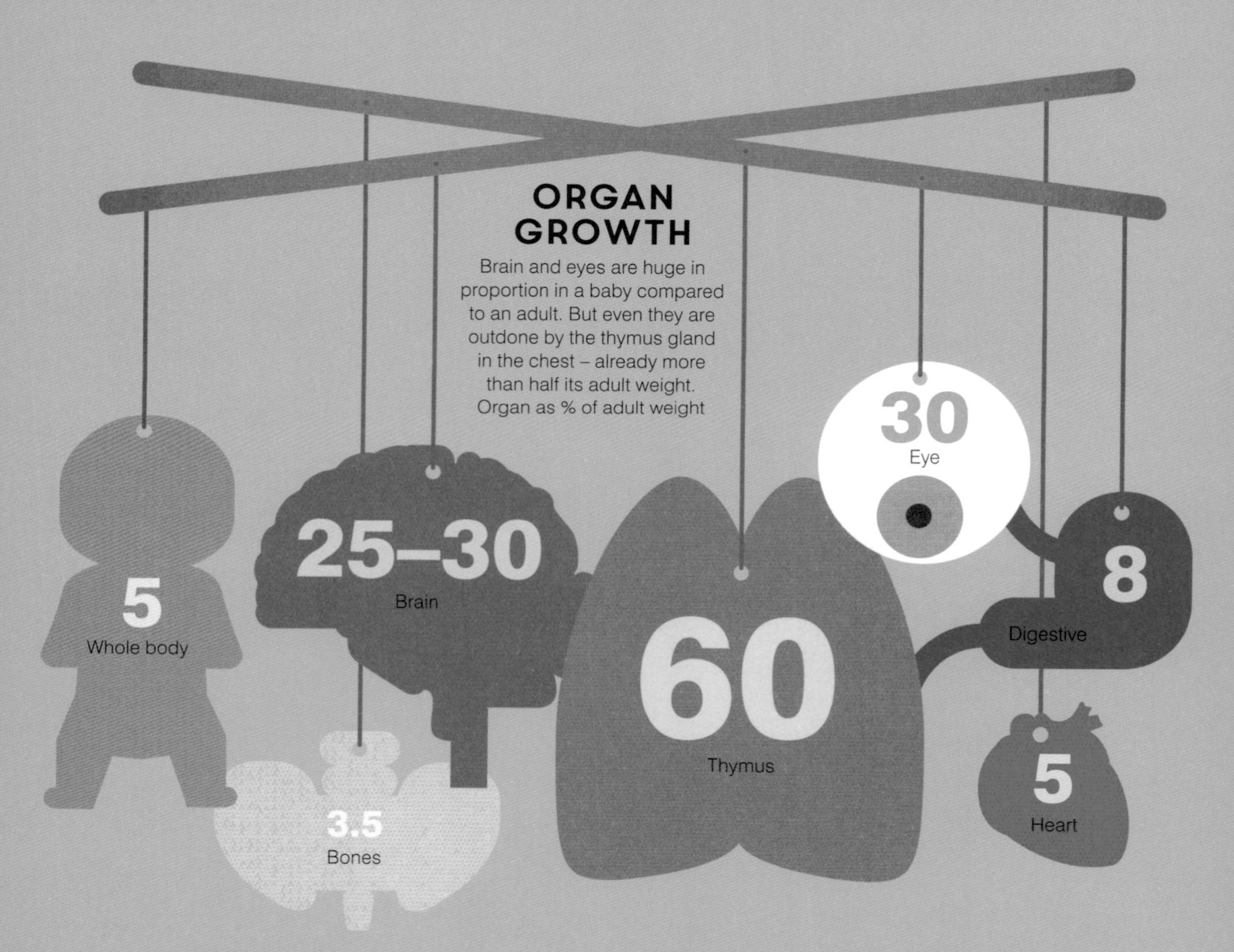

TIMETABLE OF BIRTH, FIRST-TIME MOTHERS

Overall average 12–14 hours. Second and subsequent births are usually shorter, 6–8 hours. Stages:

stages 1

Time in hours: 6–8

Phase 1: Early Uterine contractions steadily increase in strength and frequency

Phase 2: Active

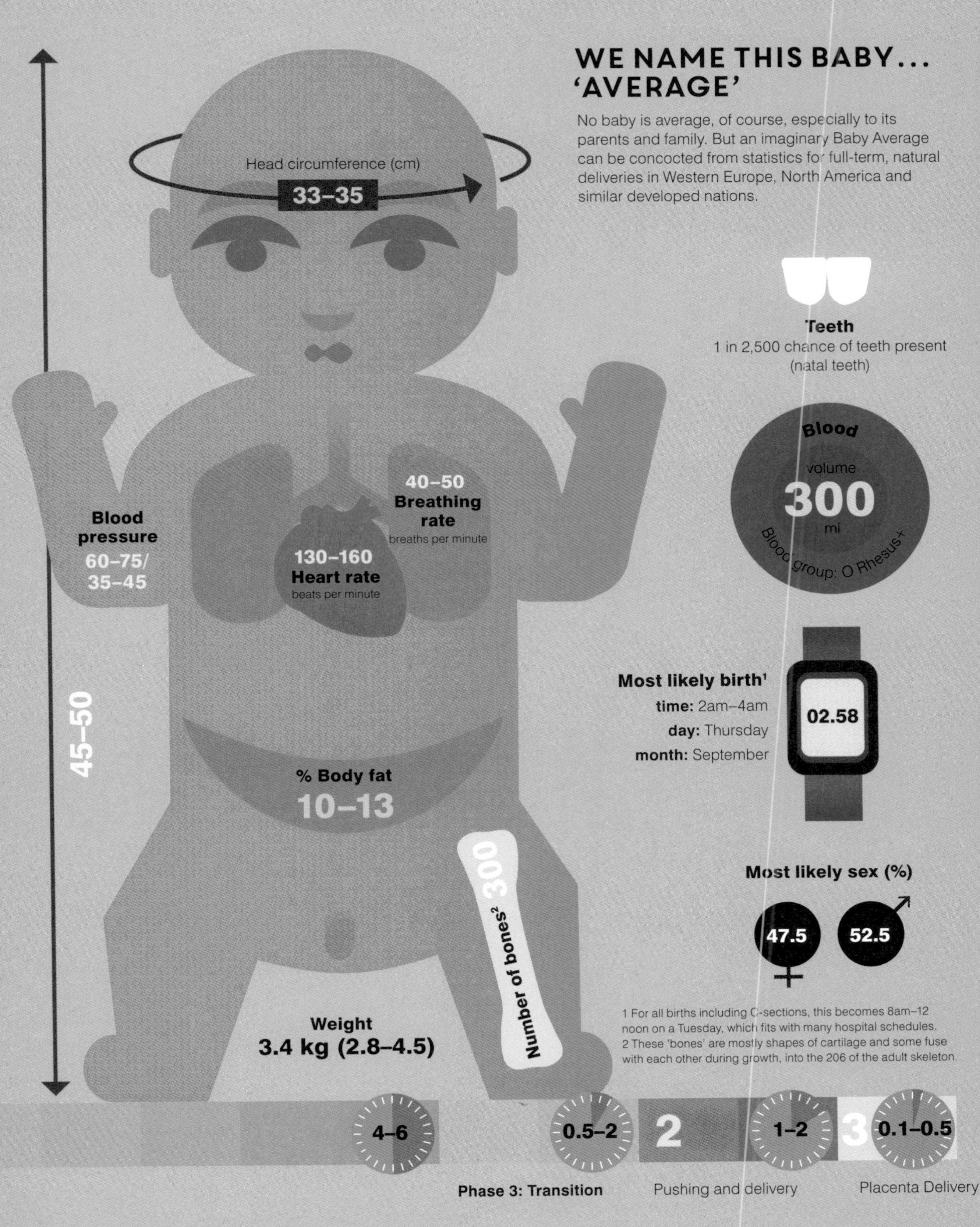
WE NAME THIS BABY... 'AVERAGE'
No baby is average, of course, especially to its parents and family. But an imaginary Baby Average can be concocted from statistics for full-term, natural deliveries in Western Europe, North America and similar developed nations.
Head circumference (cm)
33–35
45–50
Teeth
1 in 2,500 chance of teeth present (natal teeth)
Blood
volume
300
ml
Blood group: O Rhesus+
40–50
Breathing rate
breaths per minute
Blood pressure
60–75/
35–45
130–160
Heart rate
beats per minute
Most likely birth[1]
time: 2am–4am
day: Thursday
month: September
02.58
% Body fat
10–13
Number of bones[2] 300
Most likely sex (%)
47.5
52.5
Weight
3.4 kg (2.8–4.5)
1 For all births including C-sections, this becomes 8am–12 noon on a Tuesday, which fits with many hospital schedules.
2 These 'bones' are mostly shapes of cartilage and some fuse with each other during growth, into the 206 of the adult skeleton.
4–6
0.5–2
2
1–2
3
0.1–0.5
Phase 3: Transition
Pushing and delivery
Placenta Delivery

BABY TO CHILD

Each baby and child grows and develops at its own rate. Reaching one ability or skill early is not an accurate indicator of attaining others early, or eventual levels of ability. Some slow starters race ahead in later years, and vice versa. Others progress erratically. 'Milestone' times can be useful to pick up the occasional worry. Reassuringly, the vast majority of children get there eventually.

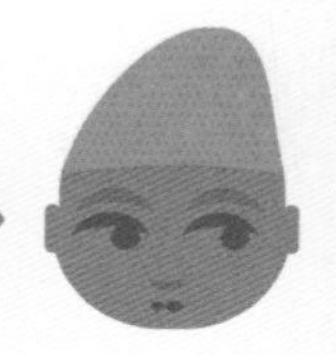

12

- Imitates movements of others
- Indicates wanting with gestures
- Says a few extra words
- Walks a few steps

15

- Vocabulary increases to 4–8 words
- Plays with ball
- Draws random simple lines
- May walk backwards with aid

18

- 'Reads' books alone
- Begins to combine words into phrases
- Scribbles expressively
- Builds simple toy brick towers

2

DOG

- Walks up steps supervised
- Names cat, dog, etc. from pictures
- Kicks a ball
- Makes short 2–3 word phrases

5

- Hops, maybe skips, swings, climbs
- Speaks full sentences with varied verbs, e.g. future and past, singular and plural
- Copies simple shapes like circle, triangle

2

- Vocalizes gurgles, coos
- Holds head up for short periods
- Eyes follow moving objects
- Smiles responsively

4

- Coos in response to talking
- Holds head up for longer periods
- Bears weight on legs
- Grasps an object

Months

9

- Combines syllables into word-like sounds
- Stands when holding onto support
- Bangs, drops and throws objects
- May mouth 'mama' type sounds

6

- Turns head towards sounds
- Rolls over in both directions
- Reaches for and mouths objects
- Sits without support

2

me me me

- Names body parts on dolls, animal toys
- Starts talking about self
- Arranges items in categories
- May start to jump

2.5

- Brushes teeth with help
- Draws lines at deliberate angles
- Puts on easy-fit clothing
- Balances briefly on one foot

Years

4

1 2 3 4

- Understands the basics of counting
- Catches a ball much of the time
- Breaks or cuts and eats own food
- Starts to copy letters when drawing

3

- Balances on one foot for several seconds
- Combines 4–6 words into sentences
- Names actions such as skip, jump, roll
- Potty-trained during day

GROWING UP

It's an amazing journey from babyhood and infancy to growing child, teenager and young adult. From birth the body increases its height between three and four times and its weight by 20 times or more. But the relative sizes of body parts at birth are far from their adult proportions, and their growth rates vary too.

GROWTH RATE CHARTS

A child on the 50th percentile means half of 100 children of the same age will be taller or heavier, and half shorter or lighter. Similarly for the 90th percentile, 10 will be taller or heavier, and 90 shorter or lighter.

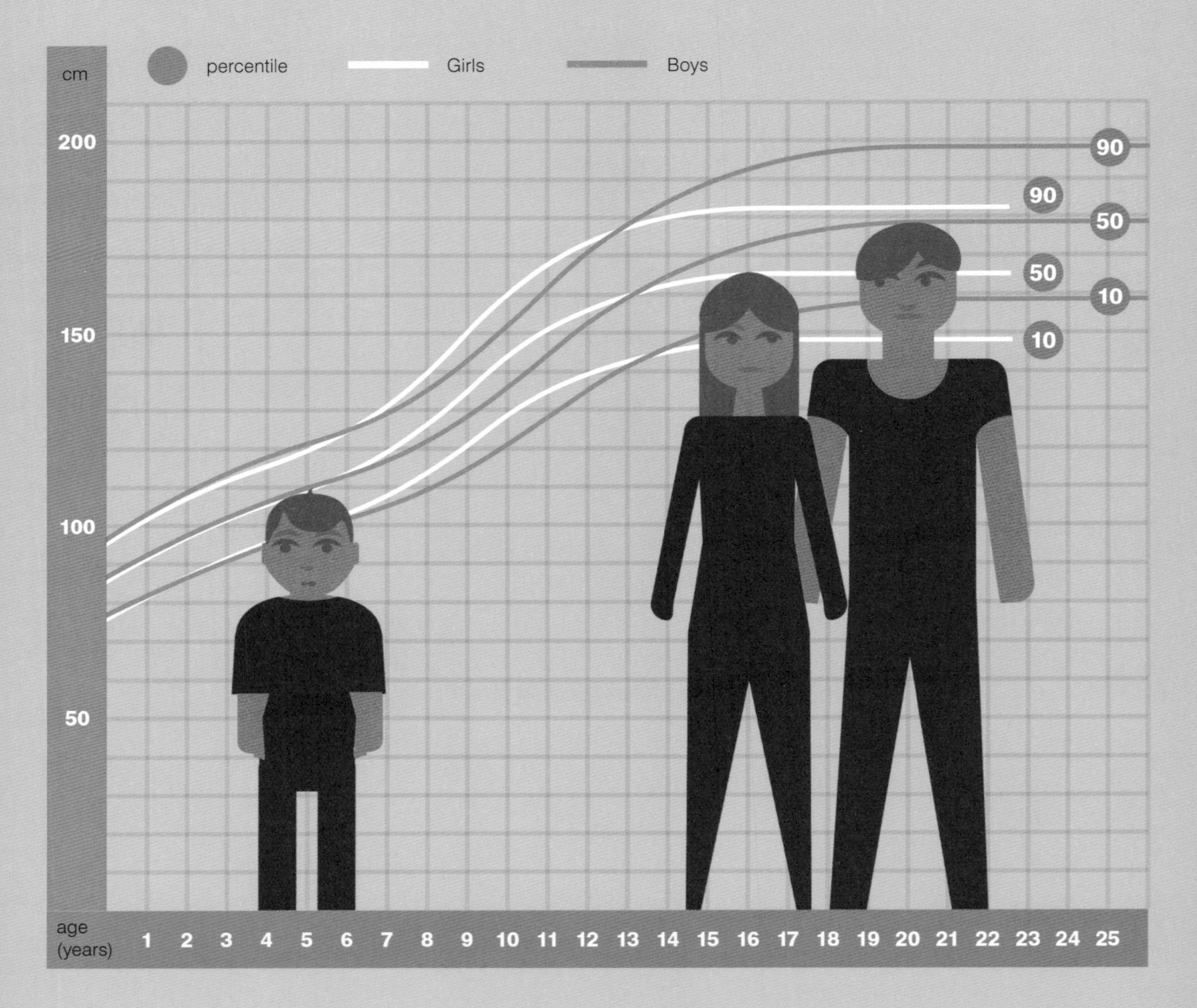

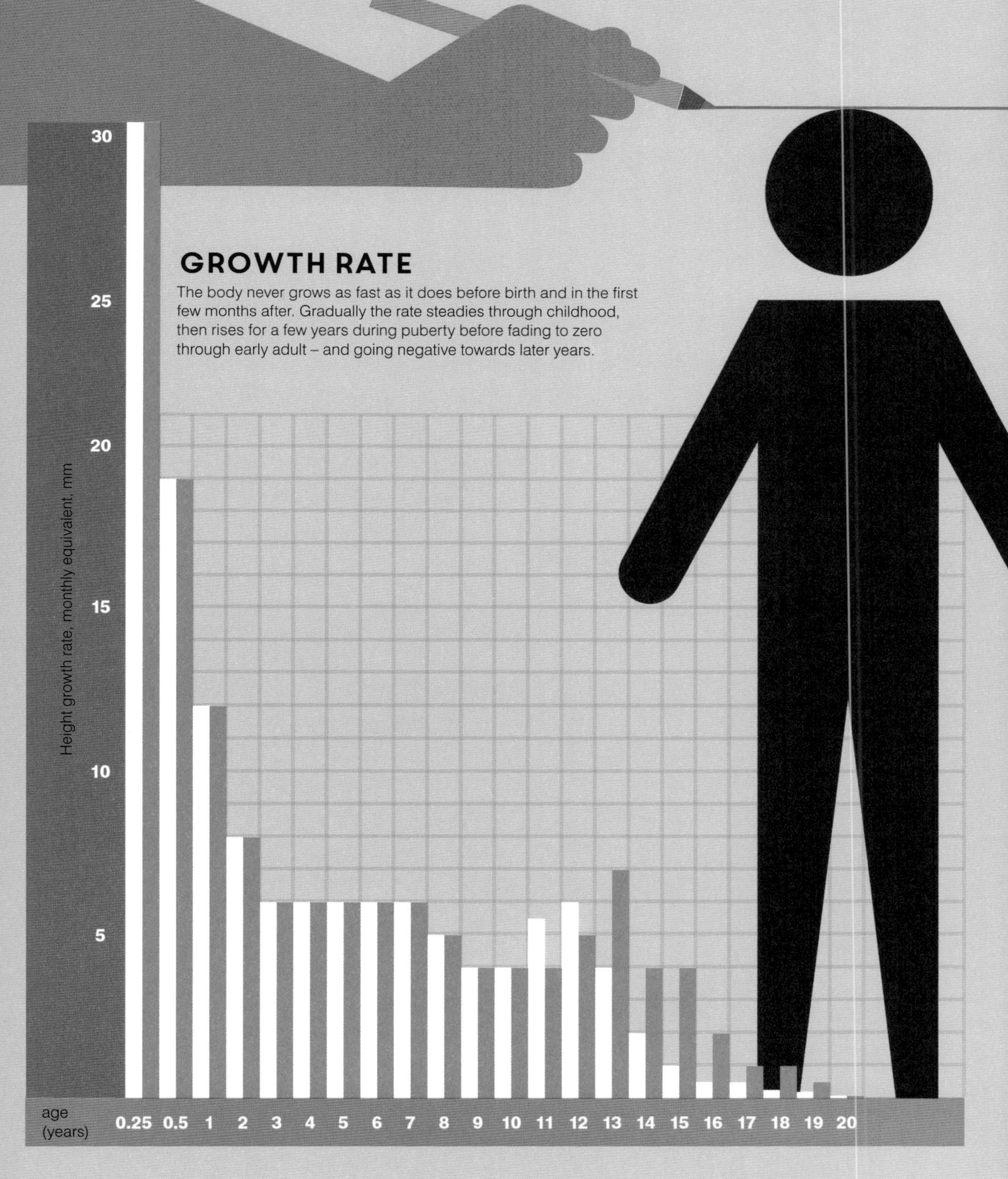
GROWTH RATE
The body never grows as fast as it does before birth and in the first few months after. Gradually the rate steadies through childhood, then rises for a few years during puberty before fading to zero through early adult – and going negative towards later years.
Height growth rate, monthly equivalent, mm
30
25
20
15
10
5
age (years)
0.25 0.5 1 2 3 4 5 6 7 8 9 10 11 12 13 14 15 16 17 18 19 20

HOW LONG DO HUMANS LIVE?

Life expectancies are complicated. Some give a 'snapshot' of estimated longevity in the general population at a certain time. Others categorize by sex and age, so women are expected to live longer than men, and estimated lifespans vary from young to old. Still others give projections for babies born on a specific date – any date. In general, all of these are getting longer. Of course, hugely important too are where people live, their health history, and – very significant – their wealth.

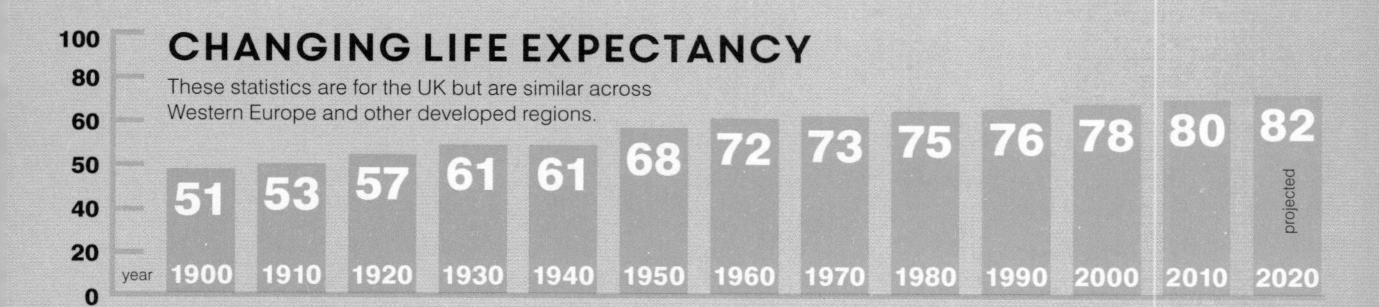

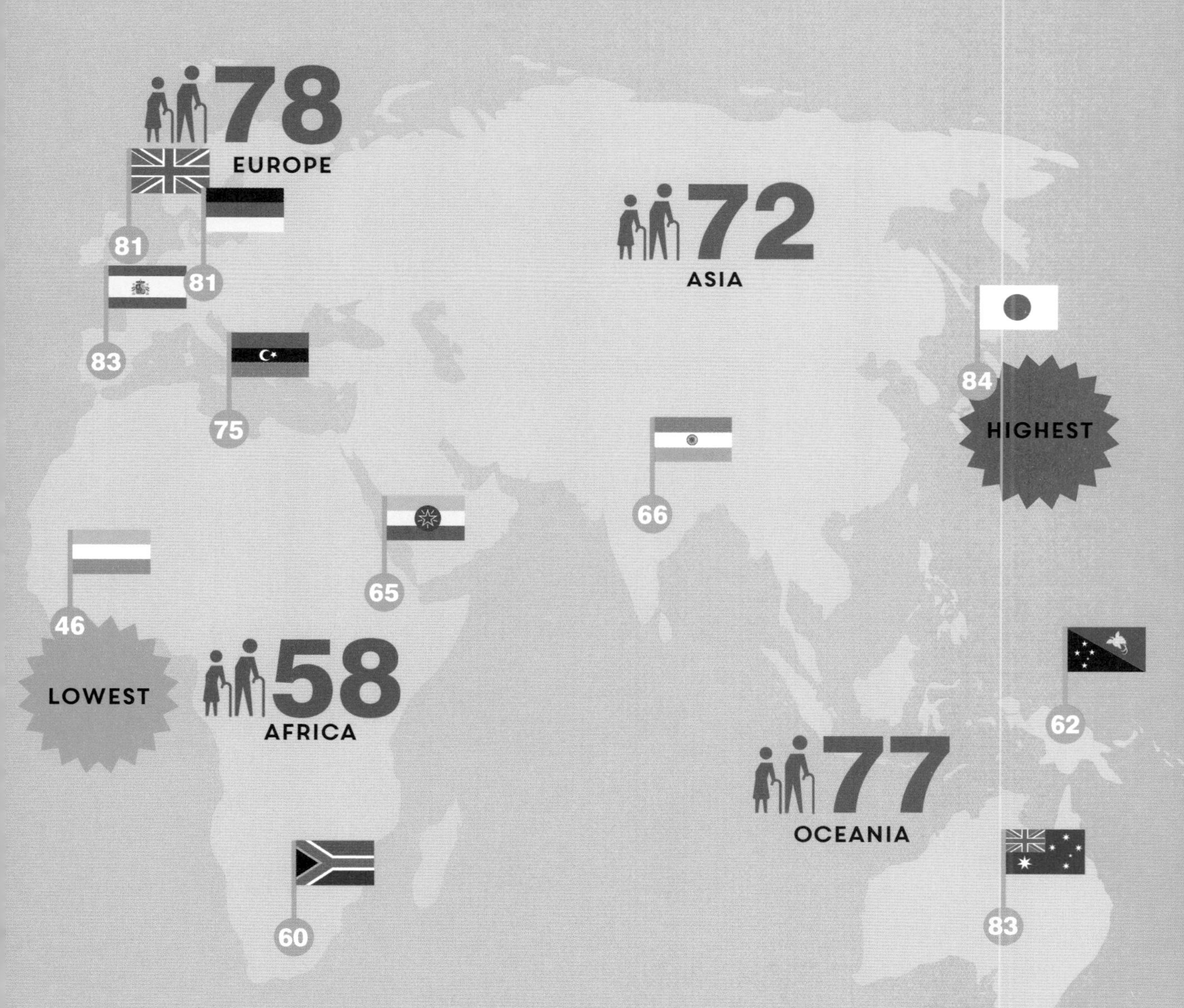

Regional life expectancy at birth, years
For babies born now, estimated from data over the last few years, since 2012.

HOW MANY NEW BODIES?

Around the world, 255 new babies are born per minute, that is, more than four every second. But this is not the global population growth rate because it is balanced by 105 deaths per minute. So the world gains an extra 150 human bodies per minute, or 210,000 per day – that's the number of people in a large city. It sounds immense, but it is lower than population growth a few decades ago.

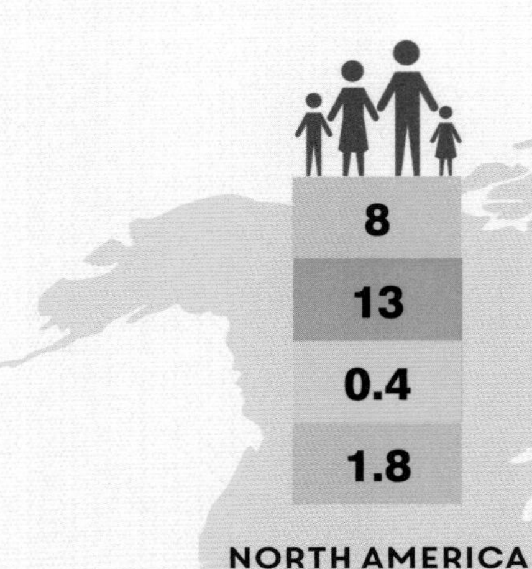

% world population

Birth rate per 1,000 people

% natural population growth, birth rate minus death rate

Fertility rate, average number of babies per mother

6

17

1.2

2.2

CENTRAL & SOUTH AMERICA

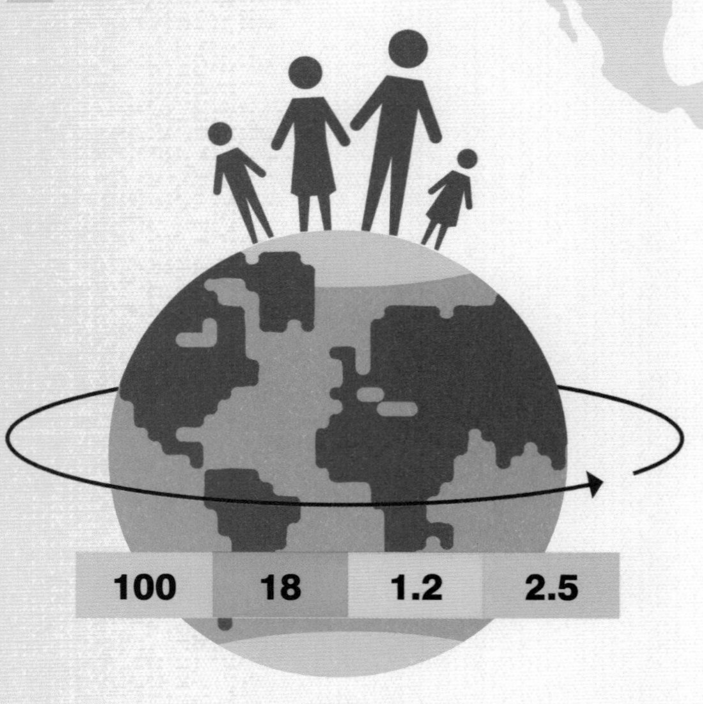

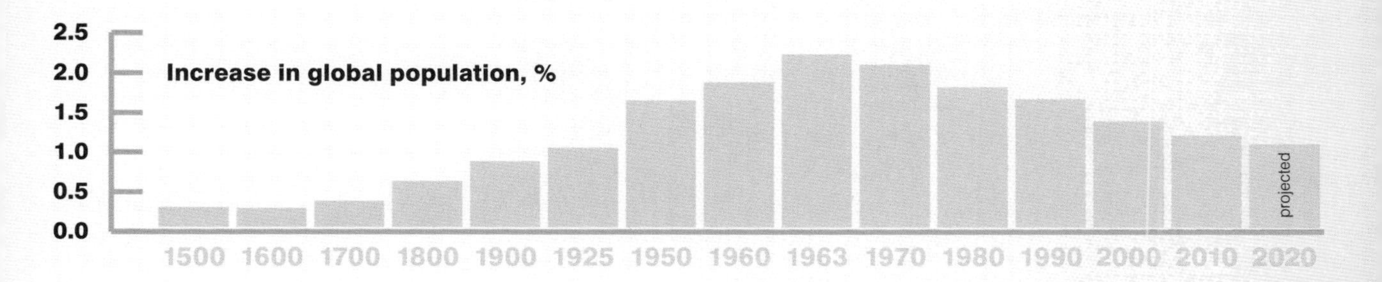

WORLD POPULATION GROWTH RATE, %

The rate at which more bodies arrive on Earth peaked in the early 1960s. In recent years the number of babies born has remained fairly constant at about 130–135 million per year. However, the growth rate is falling since, as the total rises, these babies form a reducing proportion of that total.

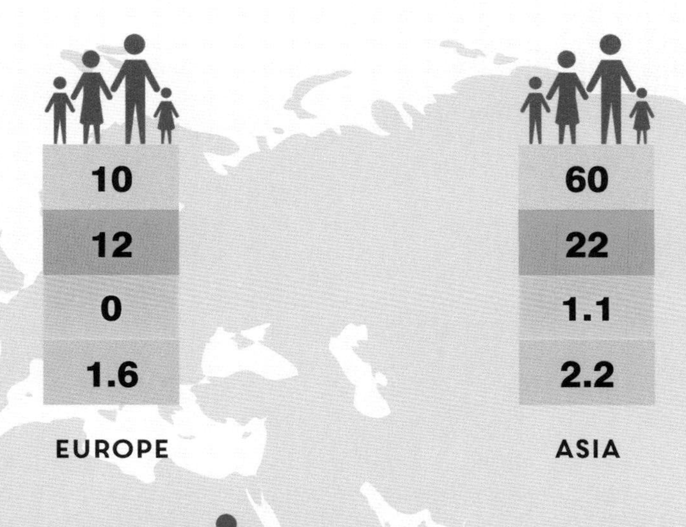

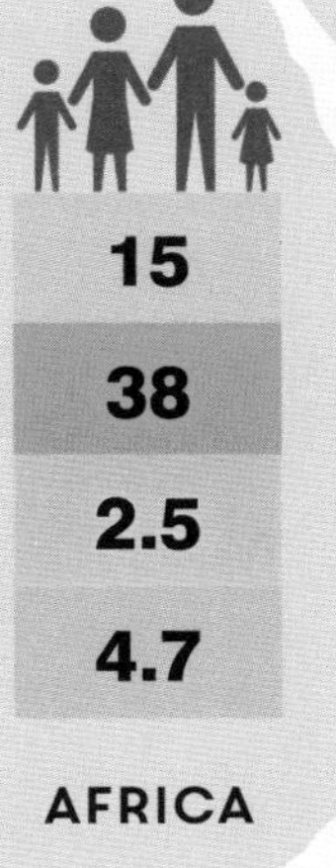

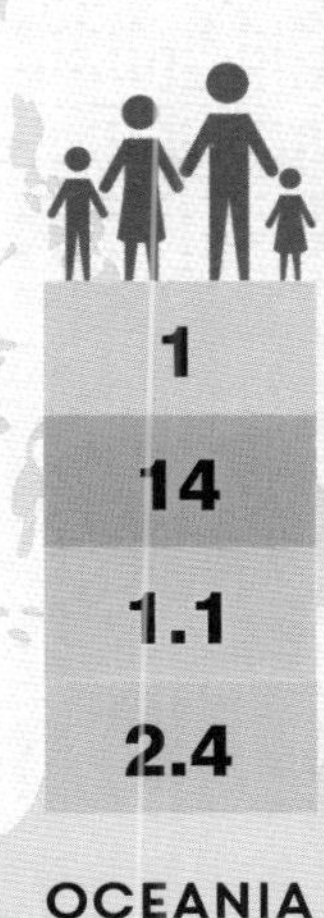

BABIES AROUND THE WORLD

Birth rates are affected by many factors, from local customs and traditions to religion, economic conditions, and government rulings such as one child per couple

HOW MANY HUMAN BODIES?

About one person in 16 who has ever lived on Earth, is alive today. Population growth in terms of number of births is steady, although the growth rate falls because they are a smaller proportion of an increasing total. Will an upper limit be reached? Many say we are already living unsustainably, and although human ingenuity for a while may continue to find quick fixes in agriculture and technology, ultimately that will cease.

GLOBAL POPULATION

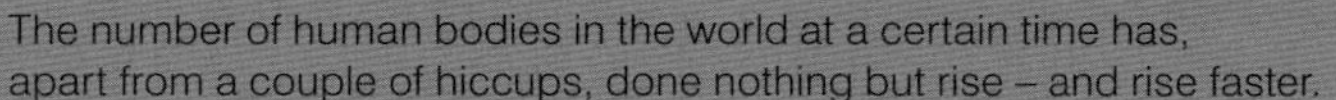

The number of human bodies in the world at a certain time has, apart from a couple of hiccups, done nothing but rise – and rise faster.

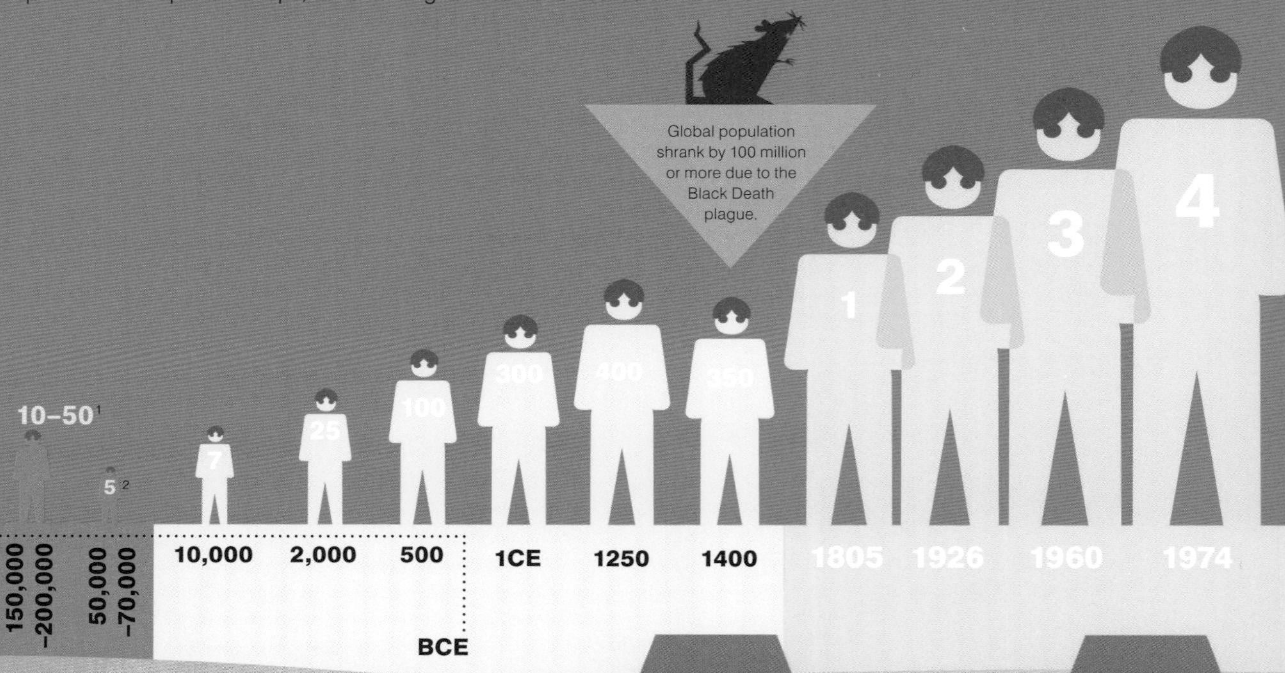

TOP CAUSES OF DEATH

in the world (recent years), numbers in millions per year

7.5
Heart disease

6.7
Stroke

1 The 'founder' population of our species *Homo sapiens* in East Africa.

2 Genetic, fossil and climate evidence suggest a 'Toba bottleneck' when modern humans (ourselves) and much other life were greatly reduced by the Toba supervolcano eruption in Sumatra.

Thousands

Millions

Billions

5

6

7

8

9

10

1987

1999

2011

2028

2045

2070

3.1

Chronic obstructive pulmonary disease

(emphysema, chronic bronchitis, etc.)

3.1

Lower respiratory infections

(pneumonia, acute bronchitis, etc.)

1.6

Lung and airway cancers

MEDICAL BODY

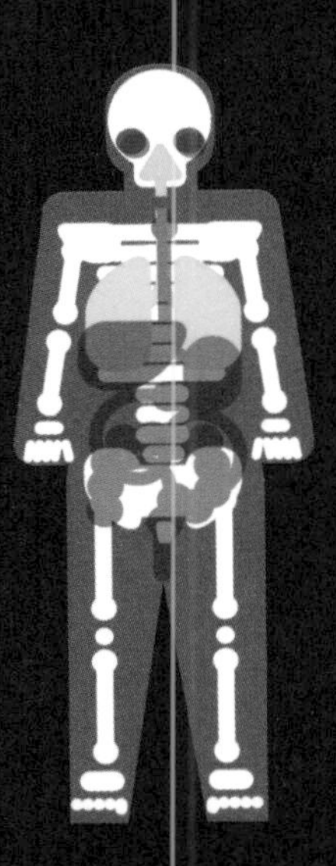

CAUSES OF ILL HEALTH

According to the World Health Organization: 'Health is a state of complete physical, mental and social well-being and not merely the absence of disease or infirmity.' There are many categories of ill health, and their causes often overlap. They can be broken down into the following groupings.

LIFESTYLE & ENVIRONMENT

Lack of exercise contributes especially to
heart disease, stroke, diabetes, cancer and depression

Smoking tobacco
is a massivecontributor to ill health

Environmental causes include
inhaled and contact toxins, infections from poor sanitation, excessive noise, disrupted routines as with shift workers, difficult social conditions

Mental problems include
stress, anxiety, depression

TUMOURS & CANCERS

When cells multiply out of control forming a growth or tumour

Benign tumours are self-contained, malignant or cancerous tumours spread or metastasize

Varied causes and triggers, from carcinogenic chemicals (e.g. tobacco smoke) to radiation (strong sunlight, X-rays), germs, poor diet

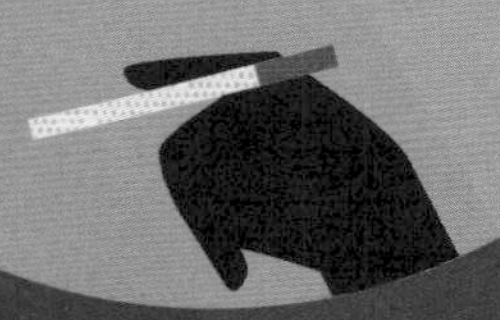

IMMUNE SYSTEM & ALLERGIES

The body's immune defence system mistakenly begins to attack its own cells and tissues, known as autoimmunity. Is a component or part of many other conditions

Examples range from
hay fever and food allergy to diabetes mellitus type 1

INFECTIONS & INFESTATIONS

Caused by germs and parasites

Main groups of germs are
bacteria, viruses and protozoa

Infectious diseases include
boils and Lyme disease (bacteria), common cold and Ebola (viruses), malaria and sleeping sickness (protozoa)

Infesting parasites include
internal roundworms, tapeworms and flukes, and external fleas, lice and ticks

INJURY & TRAUMA

May be accidental or deliberate violence.
Can occur anywhere: home, travel, work, leisure.
May result in lasting problems

DEGENERATIONS

Gradual wearing out and inadequate replacement at the levels of body cells, parts and systems

Examples include
osteoarthritis (physical joints), Alzheimer's (nerve cells), macular degeneration (eye tissues)

NUTRITION

Unhealthy or excessive diet
can contribute to obesity and many illnesses, and directly cause others

Malnutrition
leads to a host of health problems such as vitamin deficiency

Lack of hygiene and poor food preparation
may cause food poisoning

Overindulgence,
such as too much alcohol, is linked to many health problems

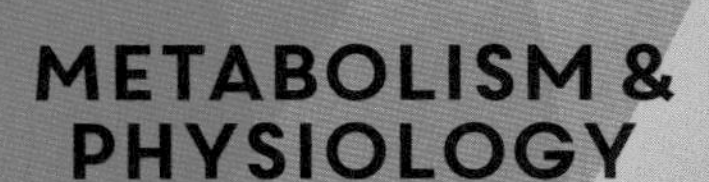

METABOLISM & PHYSIOLOGY

Problems with the body's myriad chemical processes

Causes range from dietary to genetic and environmental

Include porphyria, acidosis, haemochromatosis

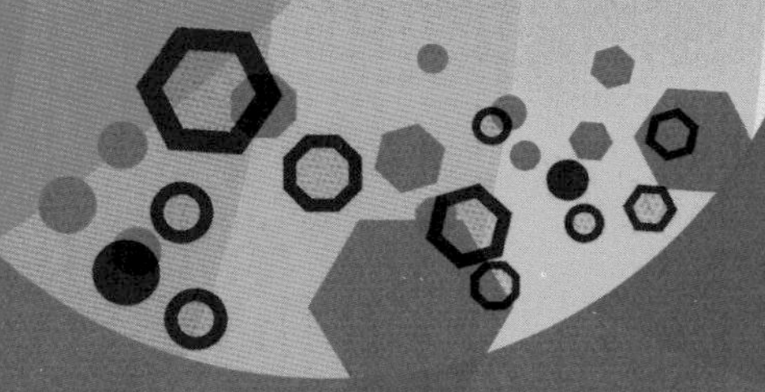

GENES & INHERITANCE

Faulty genes may be inherited or arise in the body by mutation

Some are inherited in a relatively simple way, such as sickle cell disease, cystic fibrosis

Many diseases have a less clear genetic component or tendency, such as breast cancer, schizophrenia

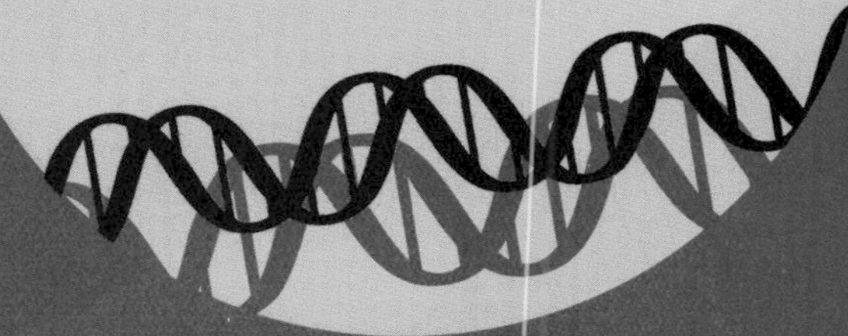

WHAT SEEMS TO BE THE TROUBLE?

Diagnosis (Dx) involves identifying or determining the nature and cause of ill health. All doctors diagnose, but for some it is a specialized field and they become expert diagnosticians. Most medics agree diagnosis is partly a science involving rational consideration of cause and effect, logical selection and elimination – and partly an art following suspicions and hunches.

PAIN IN THE ABDOMEN

The body's abdomen is packed with parts and organs. Determining the site of the pain gives clues to its origin and helps towards a diagnosis. Also important is describing the pain: dull or sharp, constant or spasmodic, burning or stabbing, related to diet or movement. The torso is therefore broken down into Quadrants and Regions for better location.

Left hypochondriac region
- Spleen abscess, enlargement, rupture
- Possible left lung or heart involvement

Umbilical region
- Small intestine, Meckel's diverticulum
- Lymph nodes, lymphoma
- Early appendicitis

Right iliac region
- Appendix, appendicitis
- Large intestine, Crohn's disease
- Ovary cyst, inflammation/infection
- Hernia

SEEING THE DOCTOR

Country	Number of doctors per 1,000 people
Japan	2.3
Germany	3.9
France	3.2
Canada	2.1
Australia	3.3
UK	2.8
USA	2.5

Number of doctors[1] per 1,000 people

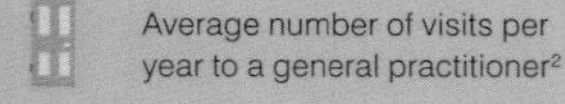
Average number of visits per year to a general practitioner[2]

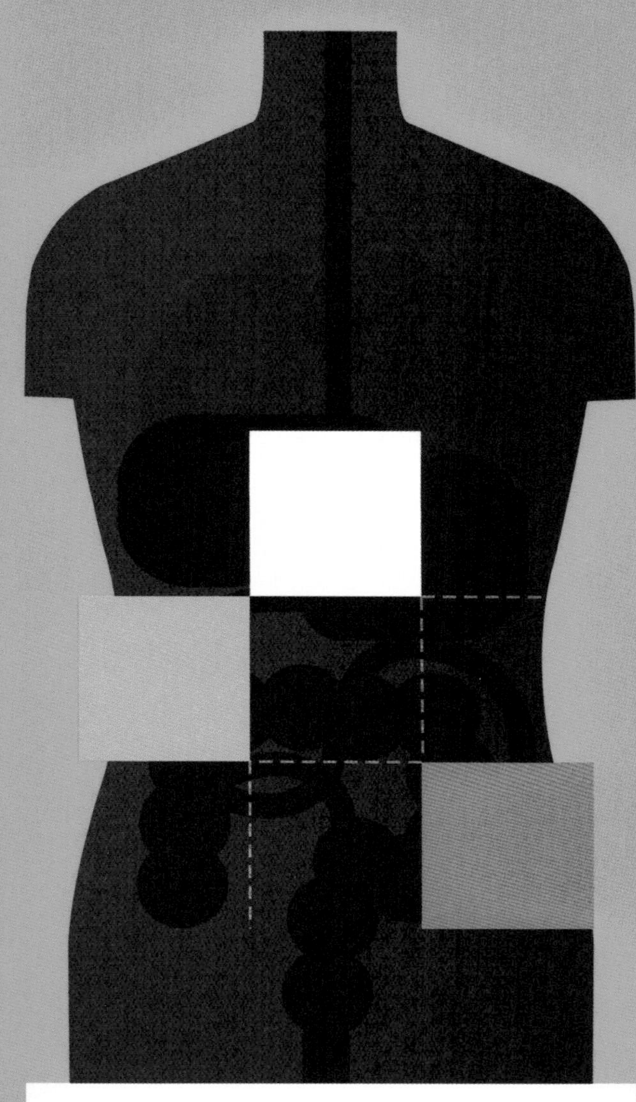

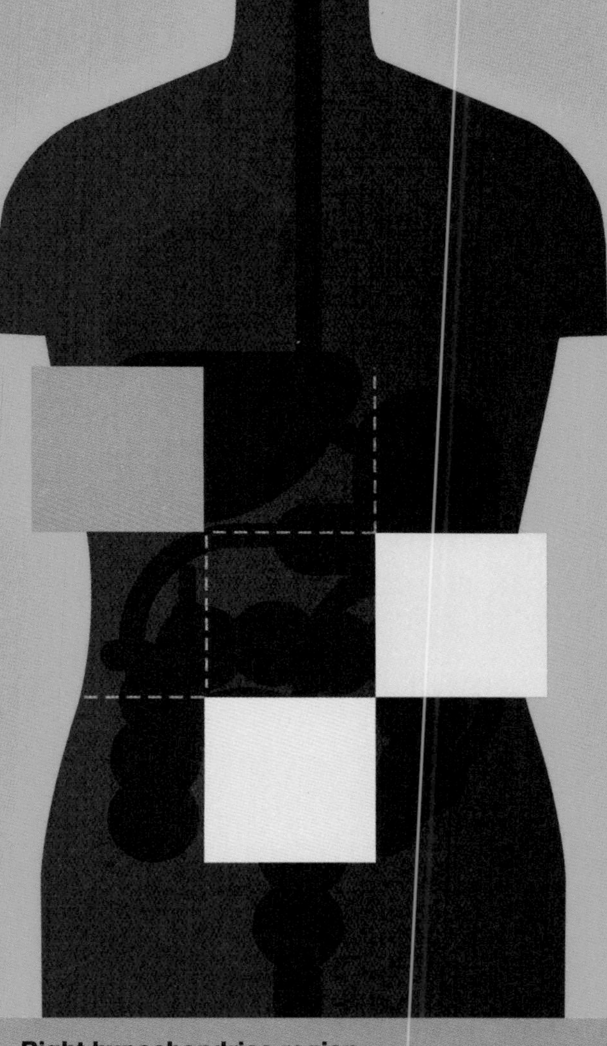

Epigastric region
- Gullet oesophagitis, stricture
- Stomach inflammation (gastritis), ulcer, gas, food toxins
- Pancreas inflammation (pancreatitis)

Right lumbar region
- Right kidney inflammation, infection, (pyelonephritis)
- Ureteric colic (kidney stone trapped in ureter)

Left iliac region
- Large intestine ulcerative colitis, diverticulitis, constipation
- Ovary cyst, inflammation/infection
- Hernia

Right hypochondriac region
- Liver hepatitis, abscess
- Gall bladder inflammation (cholecystitis), gallstones
- Possible right lung or heart involvement

Left lumbar region
- Left kidney inflammation, infection (pyelonephritis)
- Ureteric colic (kidney stone trapped in ureter)

Hypogastric region
- Bladder cystitis, stones, urine retention

1 All officially qualified medical doctors.
2 Officially qualified primary care medical doctor. Visits tend to be greater in nations with a higher proportion of older people.

MEDICAL INVESTIGATIONS

The discovery of X-rays in 1895 opened up an amazing new world of non-invasive medical imaging. Measuring the heart's electrical pulses, ECG, was developed soon after, in 1901. Today more than a dozen X-ray and scanning methods diagnose all manner of problems, from a swallowed paperclip to narrowed arteries or tumour growths. And the principle of the ECG has extended to the brain, eye and other organs.

RADIATION EXPOSURE

Almost as soon as X-rays were discovered, their harmful effects became known. In most regions, regulations limit the amount or dose of X-ray radiation that patients receive (and to which staff are regularly exposed).

μSv = microsievert, a measure of radiation dosage

0.1–1 Airport scanner
3,000 Average yearly environmental exposure
20,000–30,000 Full body CT

Nuclear scan
CT Scan
Coronary angiogram
X-Ray

Brain **2,000**
EEG Electroencephalogram brain **0.1**
ERG Electroretinogram eye, retina **0.5**
EOG Electrooculogram eyeball, eye muscles **0.1-1**
Dental **5**
Thyroid **4,800**
blood vessels, breast, pelvis **5,000–7,000**
Heart **16,000**
Mammogram **400**
ECG Electrocardiogram heart **1-2**
arm **10**
chest **100**
EGG Electrogastrogram Stomach **0.005–0.01**

ELECTRICAL RECORDINGS[1]

Sensor pads or contacts on the body's surface detect the natural tiny electrical pulses given off by the brain, nerves, heart and other parts.

Typical voltages mV[2, 3]

ULTRASOUND

Sound waves that are too high for our ears to hear are known as ultrasound. They can be adjusted to image different body parts.

1 kHz = kilohertz = 1,000 sound waves per second

10 Upper limit human hearing, older
20 Upper limit human hearing, younger
60 Upper limit dog hearing
200 Upper limit bat hearing
2,500-15,000 Medical ultrasound

Typical wave lengths kHz[1]

MRI

Magnetic resonance imaging uses extremely powerful magnets to align bits of atoms in the body.

Tesla is a unit of magnetic strength, or more technically, magnetic flux density, one weber per meter squared (one kilogram per second squared per ampere).

0.00005 Earth's natural magnetic field
0.005 Fridge magnet
1 Scrapyard recycling magnet
1.5–3 Typical MRI scanner (humans)
7–15 Powerful MRI scanner (animals)
50+ Scientific research magnets

Abdomen pelvis **15,000**

abdomen, baby **2,500–3,500**

EMG Electromyogram skeletal muscles **0.05–30**

muscles, superficial tissues **10,000–15,000**

EDA Electrodermal activity[4] **skin n/a**

Advances in non-invasive body imaging

X-rays 1895

Contrast X-rays 1896

Electro-cardiogram 1901

Ultrasound 1949

C(A)T computerized (axial) tomograph 1972

PET positron emission tomography 1973

MRI 1977

1 ****–gram is the image, display or recording produced, ****–graph is the machine that produces it, ****–graphy is the procedure.

2 mV = millivolts = 0.001 or 1/1000th volt.

3 Many of these devices measure voltage change, rather than volts produced.

4 Includes GSR, galvanic skin response. Measures how well the skin conducts electricity, rather than how much electricity it generates, as used in polygraph or 'lie detector' devices.

SURGICAL MEDICINE

Surgery – physically manipulating and altering the body – is no longer limited to 'under the knife'. Injections, chemicals, lasers and many other procedures can be involved. Surgical rates differ vastly around the world and to some extent reflect the health problems and age structure of each nation, as well as the standards of health and medical care. For example, liposuction (fat removal) tends to occur most in rich nations, while cataract procedures are relatively more common with ageing populations.

HOW MUCH SURGERY?

Proportion of people having one or more surgical procedures in one year.

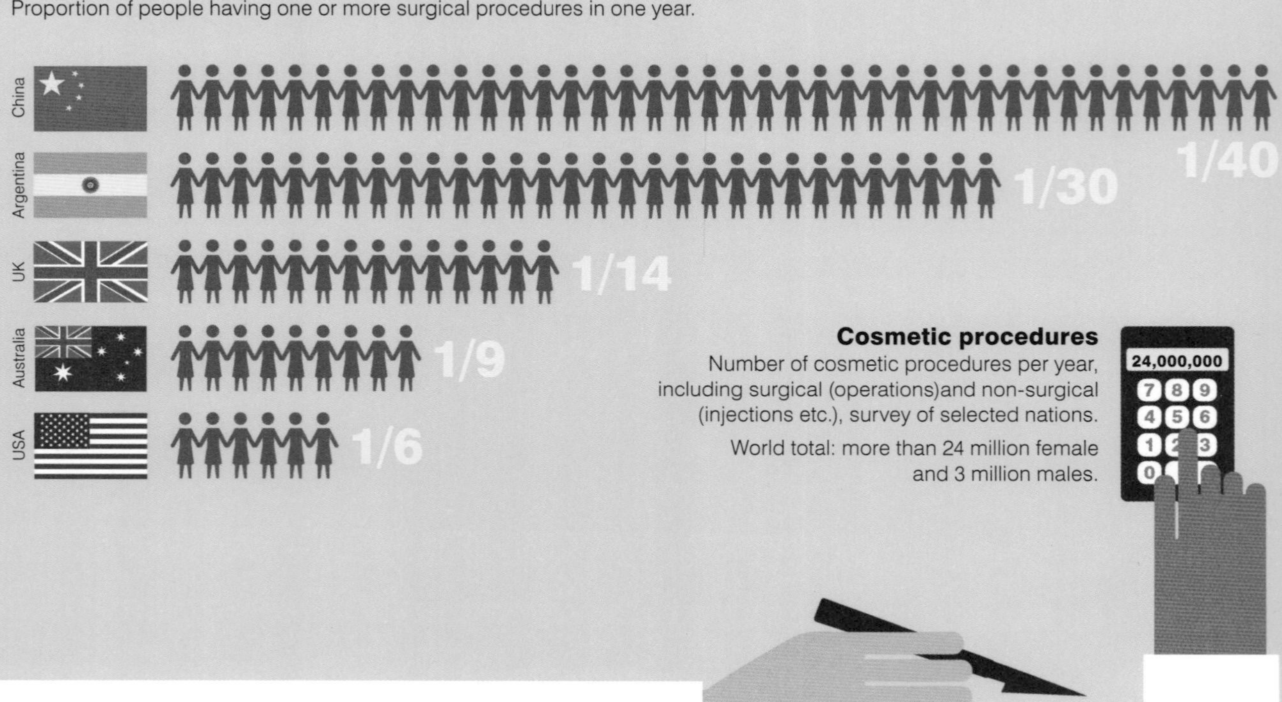

Cosmetic procedures

Number of cosmetic procedures per year, including surgical (operations)and non-surgical (injections etc.), survey of selected nations.

World total: more than 24 million female and 3 million males.

TOP FIVE COSMETIC SURGICAL PROCEDURES % OF TOTAL

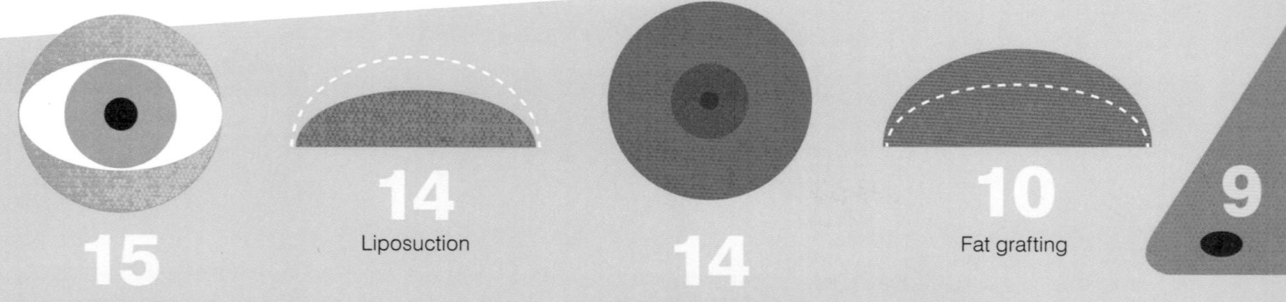